THE RESTITUTION
OF MAN

C.S. Lewis and The Case
Against Scientism

Michael D. Aeschliman

WILLIAM B. EERDMANS PUBLISHING COMPANY
GRAND RAPIDS, MICHIGAN

In piam memoriam
Adrien Rene Aeschliman
1899-1981
sapiens et eloquens pietas

Library of Congress Cataloging in Publication Data

Aeschliman, Michael D.
The restitution of Man.

1. Lewis, C. S. (Clive Staples), 1898-1963—Philosophy. 2. Scientism in
literature. 3. Metaphysics in literature. 4. Materialism in literature.
5. Philosophy in literature. I. Title.
PR6023.E926Z56   1983   828'.91209   83-8921
ISBN 0-8028-1950-8

Contents

Foreword

*T*HERE has always been a certain mystery about C. S. Lewis arising out of the disparity between C. S. Lewis the man and C. S. Lewis the Christian evangelist. He was so essentially a don living among other dons—in his ways, in his thoughts and jokes, in his teaching and style as a lecturer. At the same time, he proved to be one of the most effective of Christian evangelists, with an appeal by no means limited to any particular class, temperament, or denomination. His radio talks were frequently repeated, attracting exceptionally large audiences, and his religious writings generally have gone on being enormously popular among the erudite and the simple alike, with sales running into millions rather than thousands. A book of his like *Screwtape Letters* is now as established as, say, Swift's *Gulliver's Travels* or Orwell's *Animal Farm*, and there must be few English-speaking Christian converts of recent decades who will not acknowledge their obligation to, particulary, his *Mere Christianity*, and quote with approval and zest his words, his ideas, and his fantasies—for instance, his *Chronicles of Narnia*, now made into a successful film.

I must confess that I have myself laboured under a certain malaise in trying to appreciate Lewis as the number-one Christian apologist of our time, and to recommend him to others as such. Here, Michael Aeschliman's study of Lewis has been a great help, and I am duly grateful. With notable insight, careful scholarship, and lavish quotations, he makes the reader see that Lewis's donnish tricks and mannerisms are no more than a facade behind which there is a humble seeker after truth, a common or garden pilgrim along the dusty highways of this world. As a pilgrim, that is to say, Lewis is Bunyan's man rather than St. Thomas Aquinas's, or

John Henry Newman's, or even Cervantes's; in the Bedford tinker he rightly sees the true and lively image of Everyman making his way to the Celestial City, coping with the intervening hazards and pitfalls and eschewing all shortcuts. As Bunyan puts it,

> Some also have wished that the nearest way to their father's house were here, that they might be troubled no more with either hills or mountains to go over, but the way is the way, and there is an end.

The don echoes and loves the tinker's certainties.

It is this basically commonsensical, down-to-earth presentation of the Christian faith which, as Aeschliman indicates, makes Lewis's evangelism so appealing. There is little mysticism, in the ordinary sense of the word, in his books and addresses; he cites no vision to bolster up his faith and is not given to speaking with tongues to expound it. In the past he sees Dr. Johnson rather than one of the saints—St. Francis of Assisi, say, or St. Teresa of Avila—as the prototype of the good Christian; in his own time, not T. S. Eliot or Pope John XXIII, but the journalist G. K. Chesterton, who managed to inject the Holy Ghost, the Comforter, into the outpouring of words requisite in his profession—as it were, a gargoyle writing the script for a steeple. Or in Aeschliman's words, "As Dr. Johnson fought the impiously excessive rationalism of the eighteenth century, Chesterton and Lewis fought the excessive naturalism that has pervaded—and blighted—much of the twentieth century. For both writers, satirical essay, romance, and prose apologetic were their main literary forms and their enduring legacies."

In the great battle that has raged since the so-called Enlightenment in the eighteenth century between knowledge—or, in contemporary jargon, scientism—and faith, between *scio* and *credo*, Lewis is manifestly on the side of faith. So was Blake, who scribbled on the title page of his copy of Bacon's *Essays*: "Good advice for Satan's Kingdom." In our twentieth century more than ever, we have to sign up on one side or the other, and in making this momentous choice Lewis is a great help; the more so if we have Aeschliman's book along with Lewis's *oeuvre* as a sort of glorified footnote. I like very much Aeschliman's last word about the man whose thoughts and beliefs he has studied so conscientiously, analysed so minutely, and explained so coherently: "He [Lewis] bowed to the constraint of truth and thus climbed its ladder."

Malcom Muggeridge

I

Common Sense and the Common Man

If the ordinary man may not discuss existence, why should he be asked to conduct it?

G. K. Chesterton

. . . *common* may also contrast the *sensus* of humanity in general, favourably, with what is thought of or felt by the irrational, the depraved, the sub-human. *Common,* so taken . . . is the *quod semper quod ubique,* the normal and indeed the norm.

C. S. Lewis,
Studies in Words

SINCE the middle 1940s when he made his BBC radio talks, Clive Staples Lewis has been one of the best-selling authors in the English language. Such popularity is an honor rarely accorded a scholarly man, and especially one whose works are for the most part so thoroughly and explicitly concerned with propagating religion in a manner neither scholarly nor specialized. Since his death in 1963 after a career spent teaching Medieval and Renaissance literature at Oxford and Cambridge universities, purchases of his books have increased sixfold to more than two million volumes annually in Britain and America. Such figures substantiate the assertion of an American scholar that Lewis has had "an impact . . . on the American religious imagination . . . rarely, if ever, equalled by any other modern writer."[1] It is worth asking why and how works by such a naturally private and scholarly man should have reached and held such a vast audience.

The lapsed Catholic writer James Joyce treasured a remark that a friend once made about St. Thomas Aquinas: "The difficulty about Aquinas is that what he says is so like what the man in the street says." Later, when living in Paris, Joyce was discussing Aquinas's theology in a café when someone objected that it was all irrelevant, that it "has nothing to do with us," to which Joyce replied, "It has everything to do with us."[2] No two writers could be more different in intent or method than Joyce and Lewis, but it is the "common sense" of Aquinas that delighted both of them.

The late Basil Willey said of Aquinas that "His assumption is that religion is rational, and Reason religious,"[3] but it is not of course only of Aquinas that this is true; it is the legacy of a rational Theism of great antiquity variously fostered by Jews, Moslems, Christians, and virtuous pagans, and which a translator of Aeschylus has called "that religion which has existed since the beginning of the world, but is now called Christianity."[4] The durable conception of *recta ratio* ("right reason") is that reason rightly used—sincerely, consistently, honestly used—leads naturally and inevitably to God. It is the great central philosophical/metaphysical tradition of the West in which Lewis enlisted his own mind and pen.

The great Muslim thinker Al-Ghazali said that reason was "God's scale on earth,"[5] thereby giving expression to a tenet that reverberates throughout Roman, Greek, Jewish, and Christian traditions as well. The sources on whom Lewis chiefly relied are Plato and Aristotle, St. John, and Augustine among the Ancients, and Aquinas, Richard Hooker, Samuel Johnson, and G. K. Chesterton among the Moderns—all exemplars of what he calls "the great central tradition." Lewis loved Hooker, through whose book *On the Laws of Ecclesiastical Polity* the Delphic injunction "nothing to excess" and the Aristotelian doctrine of the Golden Mean were chiefly and widely disseminated in England. With Hooker, Lewis believed that "[t]he general and perpetual voice of man is as the sentence of God himself,"[6] but he did not confuse this tradition with the idea that majorities are always right *(vox populi vox Dei est)*. What Lewis trusted was the fund of "common sense" of men throughout history, the *communis sensus* of Vincent of Lerins, what Alexander Pope was referring to when he wrote that "whatever is very good sense must have common sense in all times":[7] the vast common sense of humanity, of which he felt he was a trustee and which he articulated and defended in all of his writing, speaking, and living.

Part of the reason for Lewis's popularity is his assumption that almost all good men who have ever thought honestly share universal convictions which may differ in detail but not in substance. He felt that the amorality, agnosticism, and atheism of much of twentieth-century culture, and especially of the culture of modernism, amounted to an aberration within the historical tradition of common sense, and that its adherents were, in the terms of Augustine whom he quotes, "divorced by some madness from the *communis sensus* of man."[8] His own idea of the perennial philosophy was that "good is indeed something objective, and reason the organ whereby it is apprehended."[9] Like Dr. Johnson he knew and tirelessly repeated the fact that most men, whatever they say or write, assume in practical conduct that they have free will and decision and that objective truth and value exist. As the American philosopher William Barrett has recently written, "We still go about our everyday business guided by this moral will, and we still discriminate on its terms. . . . In short, without being aware of it, we do follow Kant's view that the moral will is the center of the personality. And yet, amazingly enough, modern philosophers have yet to come to terms with this fact."[10] It was a fact that Lewis never forgot.

Doubtless the confidence, depth, and conviction of Lewis's writing is attributable in part to his having been a brilliant student of philosophy at Oxford before becoming a scholar and teacher of English literature. But it is crucial to note that his great learning was always worn very lightly and always subordinated to common sense, which was for him, as for Chesterton, the "metaphysical journalist" whom he so admired, something both luminously intelligible and joyously mystical. It is in his knowledge of, trust in, and dogged appeal to the common sense of mankind, the *consensus gentium*, and in his obstinate opposition to what he considered the merely eccentric, "original," or fashionable, however popular for the moment, that much of Lewis's enduring power and attraction as a thinker and a writer lie. He wrote with both delight and reverence of "the real common will and common reason of humanity, alive, and growing like a tree, and branching out, as the situation varies, into ever new beauties and dignities of application."[11]

With a profound historical imagination like that of Burke, Lewis knew, as did the medieval Bernardus Sylvestris, that if on some issues we know more than the Ancients, "we see farther because we stand on the shoulders of giants."[12] But he was also aware of the decline and loss of much essential philosophical, religious, and moral wisdom in the twentieth century—what Marcel called the "decline of wisdom"—a development in which Lewis saw the threat of destruction for man's unique nature and prerogatives, the abolition of man as a moral being, and the replacement of the image of man as *homo sapiens* with the brutally reductive and contradictory image of man "beyond freedom and dignity."

In making his case against the pervasive twentieth-century materialistic scientism apparent in cant such as Skinner's "beyond

4

freedom and dignity," Lewis made it an essential part of his strategy to employ and sharpen his wits in *vernacular* writing and speech, the defense of which was one of his chief concerns. He opposed linguistic inflation and what he called "verbicide," not out of a philologist's self-defense or scholarly snobbery but out of a commitment to truth. He dreaded the hypnotic and delusive claims and effects of "specialist" discourse, whether scientific, philosophical, or literary, seeing in the languages and pretensions of the elite the danger of a new priesthood unencumbered by traditional moral restraints: "Any fool can write *learned* language," he wrote. "The vernacular is the real test. If you can't turn your faith into it, then either you don't understand it or you don't believe it."[13]

Lewis loved Bunyan and used quotations from *The Pilgrim's Progress* in much of his writing and also imitated him in *The Pilgrim's Regress*. He had real affinities of style and content with Swift, Johnson, and Chesterton as well. A premise that all of these writers shared is that there is in the world a comprehensive and comprehensible truth, one that accommodates itself to every level of intelligence and is thus available to all men. It is also the assumption underlying the Gospel; Swift, himself a clergyman, argued in his "Letter to a Young Clergyman" that this must be true, "or else God requires of us more than we are able to perform."[14]

"Common sense to an uncommon degree," Coleridge said, "is what the world calls wisdom." This wisdom consisted largely for Lewis in a trust in and loyalty to the traditional virtues as the outline of an encompassing reality. He resented and opposed the implicit claims of both scientists and "humanists" to moral superiority over the common man. With the decline of religion among the intelligentsia, the claims for the moral importance of art and science grow; but, wrote Lewis,

> Courtesy to our contemporaries must not forbid us to point out that a poet, an admitted and unmistakable poet, is sometimes (in certain periods, often) a man inferior to the majority in 'tenderness', 'enthusiasm', and 'knowledge of human nature'—not to speak of information, common-sense, fortitude and courtesy.

Thus he opposed the excessive claims of both the cult of science, the technical professoriat that would become a new manipulative priesthood of "Conditioners," and the cult of culture, the idolatry that sought to replace religion with art and literature as guides to truth and conduct in school curricula and daily life. It is, he wrote,

> extremely doubtful whether 'culture' produces any of those qualities which will enable people to associate with one another graciously, loyally, understandingly, and with permanent delight. When Ovid said it 'softened our manners', he was flattering a barbarian king.[15]

Lewis was convinced that every person was in essence *homo sapiens*, the moral and philosophical knower, meant to live happily and considerately—happily *because* considerately—in community with his fellows, and that this was the elementary "natural" tendency with which every human creature was endowed by his Creator, however much it might be weakened or blighted by sinful egotism, by excessive claims for the self. But he was equally certain that this elementary spiritual and moral knowledge— available alike to the simple and complex, but requiring personal response to be developed into virtue, character, and the knowledge of God—was under attack from many sides. One particularly significant threat arose as a part of the climate of opinion first generated during the scientific revolution of the seventeenth century, which led to an ultimate failure of nerve and belief, one of whose effects G. M. Young describes in his *Victorian England*: "The common residual intelligence" was becoming by the end of the Victorian period "impoverished for the benefit of the specialist, the technician, and the aesthete: we . . . go out into the Waste Land of Experts, each knowing so much about so little that he can neither be contradicted nor is worth contradicting."[16] Aldous Huxley described the same problem, that "intensive specialization tends to reduce each branch of science to a condition almost approaching meaninglessness," and he added that there "are many men of science who are actually proud of this state of things. Specialized meaninglessness has come to be regarded, in certain circles, as a kind of hall-mark of true science,"[17] although it ought to be added that this is largely true of other fields too—and especially of literary criticism and philosophy.

6

In opposition to this trend toward fragmentation and de-moralization now so characteristic of modern culture, Lewis attempted to "see life steadily and see it whole," but he also insisted that it is every person's birthright and duty to do the same, and that it is in precisely this effort of humane moral and intellectual striving that much of our specifically human nature and integrity consists.

Lewis feared for the fate of common sense in the modern world, feared for the fate of the common intellectual and moral norms which are never so much an average as a standard. He believed that civilization was frail, as this century's history has shown it to be, and he opposed the corrosive skepticism of the fastidious elites who could destroy the foundations of moral order. In *The Pilgrim's Regress* he quoted Thucydides: "And every shrewd turn was exalted among men . . . and simple goodness, wherein nobility doth ever most participate, was mocked away and clean vanished";[18] and in *A Preface to Paradise Lost* he wrote,

> That elementary rectitude of human response, at which we are so ready to fling the unkind epithets of 'stock', 'crude', 'bourgeois', and 'conventional', so far from being 'given' is a delicate balance of trained habits, laboriously acquired and easily lost, on the maintenance of which depend both our virtues and our pleasures and even, perhaps, the survival of our species.[19]

Or, as Dostoevsky had put it, "humaneness is only a habit, a product of civilization. It may completely disappear."[20] The barbarities of modern history, of Auschwitz and Cambodia and the Gulag, have provided grim attestations to the vulnerability of civilization to the collapse of moral order.

To the debilitating modernism which counts nothing sacred, Lewis juxtaposed the health of vital tradition, which he strove to enrich and transmit. As the human mind can transcend time and place, it can draw on the resources of past times to correct present imbalances. Lewis praised Chesterton's essay "On Man: Heir of all the Ages," noting that an "heir is one who inherits and 'any man who is cut off from the past . . . is a man most justly disinherited.' "[21] In reappropriating, transmitting, and making newly

7

available the intellectual, artistic, and religious riches of the past, Lewis has been acknowledged as one of the great figures of our century. The scholar J. H. Miller writes of "the grand tradition of modern humanistic scholarship, the tradition of Curtius, Auerbach, Lovejoy, C. S. Lewis."[22]

The final pages of *An Experiment in Criticism* contain one of the great modern defenses of the humanities and of traditional education, explaining the meaning and importance of moral imagination and creative reading, and in so doing they provide fit company for Chesterton's brilliant essay "On Reading." Nor does Lewis restrict his comments or attention to "high culture" and the highly educated: "Those who have greatly cared for any book whatever may possibly come to care, some day, for good books," he writes. "The organs of appreciation exist in them."[23] The traditionalism that Lewis articulates is generous in its sympathies and goals, rooted in the *consensus gentium.* He has a fundamental trust in what the wisest men of the past have discovered and transmitted, and a desire not only to preserve it but to apply it and to imitate it.

As in so much else, Chesterton was influential in the development of Lewis's respect for tradition. "Modern men are not familiar with the rational arguments for tradition," Chesterton has pointed out; "but they are familiar, and almost wearily familiar, with the rational arguments for change. . . . The language which comes most readily to everyone's mind is the language of innovation; but it is a language that is rather exercised than examined."[24] Lewis examined language, literature, thought, and conduct in the light of "[a]uthority, reason, and experience" on which "in varying proportions all our knowledge depends,"[25] and in their light he found much of modern culture wanting; he profoundly mistrusted the "canonization of the primal, nonethical energies" characteristic, according to Trilling, of modern literature, though he himself warned against premature moralism in estimating literary value. While eager to extend a qualified autonomy to aesthetics and the imagination, Lewis tried in all of his own writing to "instruct by delighting," as Sidney said the poet should do.

He was in profound and life-long agreement with Dr. Johnson that "men more frequently require to be reminded than informed," and that in the essential moral sphere what they need reminding of is the "old, platitudinous, universal moral law,"[26] which he often called the "Tao," drawing attention to the fact that it was a universal fact of human experience and not a law exclusively possessed or created by any one culture.[27] So too in his aesthetic theory and practice, Lewis tried to bring to life the enduringly true; it was his conviction that no good artist, writer, thinker, or moral agent ever really creates the truth, but that in each case he or she responds to it as a preexistent and external reality, something to be touched, tasted, imitated, copied, or revived in order to "carry truth alive into the heart with conviction." There is a transcendent pattern of truth, beauty, and goodness which man by means of rational effort or inspiration is capable of knowing and experiencing. This idea is at the heart of the vision of God, the *visio Dei*, of the great theological poets Dante, Spenser, and Milton, whose work Lewis so loved; and it was part of his great effort as a scholar and teacher to make more available to modern man the richness of wisdom, beauty, and happiness to be found in the experience of reading their works.

With his emphasis on memory and reminiscence, Lewis stands in the tradition of Plato: he tries to draw out of his readers a knowledge which seems new as it occurs to us, but has the taste of reminiscence to it, as if it were something we have always known, once dimly, now more clearly. This knowledge and experience is thus always new and always old—*"tam antiqua, tam nova,"* in Augustine's words—and it is "common knowledge" in the exalted sense, part of the humane legacy that we are uniquely fitted and free as creatures to possess. In the abstract it consists largely of the enduringly valid qualities and virtues of humane living, but as Lewis notes in a short introduction to Spenser,

> What looks like a platitude when it is set out in the abstract may become a different sort of thing when it puts on flesh and blood in the story; according to the theory which Sidney set out in his *Defence of Poesie*, the poetic art existed for the precise purpose of thus turning dead truism into vital experience.[28]

Working within this great aesthetic tradition of intelligible beauty, Lewis seeks to give us the opportunity to "taste the truth of the truism"; and so when "truth hath lost her lustre," he tries to

9

burnish it anew so that what Aquinas termed its "effulgent radiance" can shine forth and be perceived as such. This is one of Lewis's chief goals, not only in the children's novels, but also in the space trilogy, in the literary criticism, and in all of the apologetics, in which his frequent satire is intended to smash partial, false, and delusive idols of truth and well-being so that the true, uncreated image of truth, available to all men, can shine forth and be newly apprehended.

His friend Dom Bede Griffiths has suggested that Lewis's powers "did not reach the supreme fulfillment in the poetry of prophecy, as one might have hoped,"[29] and this is true: he never produced the great epic to justify God's ways to man, as Dante, Spenser, and Milton had. Perhaps only Eliot among the moderns has really come close to this range of poetic achievement in English; the times, shot through with heresies and aesthetic depression, have not encouraged or supplied the context for the visionary poet. Lewis remarked in 1936 that the power of painting virtue convincingly was absolutely unknown in his generation.[30] As Dr. Johnson fought the impiously excessive rationalism of the eighteenth century, Chesterton and Lewis fought the excessive naturalism that has pervaded—and blighted—much of the twentieth century. For both writers, satirical essay, romance, and prose apologetic were their main literary forms and their enduring legacies.

Lewis's extraordinarily eloquent body of writing invites readers to "bend to the constraint of truth," and by so doing to climb the ladder of vision. In his emphasis on the reader's experience of truth, goodness, and happiness in reading good or great books, Lewis was arguing against the fashionable currents of literary criticism as they reigned during much of his career. He defended the common reader and his potential for growth and even of beatitude. "To read Spenser," he once wrote, "is to grow in mental health."[31]

Lewis's critical emphases have proved fruitful in the long run, having helped reclaim the riches and resources of Milton and Spenser especially; some of the best literary scholarship has in recent years followed explicitly the path he opened. In one of the finest books on Milton in recent years, a young scholar not only employs Lewis's insights and priorities but also reaches nearly identical conclusions about the possibilities for growth and experience that the greatest literature makes available. Discussing "the invitation to ascend" in *Paradise Lost*, Stanley Fish writes,

> The arc of the narrative describes a Platonic ascent, which culminates (for the reader who is able to move with it) in the simultaneous apprehension of the absolute form of the Good and the Beautiful, 'without shape or colour, intangible, visible only to reason, the soul's pilot'.[32]

This is the aesthetic method—and the doctrine—of the intelligible Good, the *visio Dei*, available to every individual, to which Lewis always referred and which he tried imaginatively to embody to some extent in all his writing.

He was convinced that the experience of the intelligible Good is not only available but obligatory to every person at some level and in some sphere, and that it has never been the possession merely of mystics or adepts or specialists. It is part of the common wealth of humanity, this invitation and this command to ascend and transcend the self, and to the extent that we accept the invitation and respond to the command, we are truly human, truly *homo sapiens*, and not merely *homo sciens*.

Among the alternative visions of man's essential nature and purpose, one that Lewis particularly opposed was the view of modern naturalism exemplified by a statement in Jacob Bronowski's *The Identity of Man*: "man is a part of nature, in the same sense that a stone is, or a cactus, or a camel."[33] This is a view that Chesterton abhorred as well, and that he attacked in a book that Lewis loved and often recommended, *The Everlasting Man*:

> It is not natural to see man as a natural product. It is not common sense to call man a common object of the country or the seashore. It is not seeing him straight to see him as an animal. It is not sane. It sins against the light; against that broad daylight of proportion which is the principle of all reality.[34]

The distinction between these two senses of *nature* (i.e., *nature* as the essential character of a person, reality, or thing and *nature* as the vast mechanism of the physical universe) has served to generate much of the intellectual and moral history of Western culture during the past four hundred years. The chief fallacy of modern naturalism is its insistence upon a collapse of the two meanings of *nature* into one: for Bronowski and others like him, the essential character of man amounts to nothing more than his physical existence within the larger soulless ticking mechanism of the "natural" world, constituted by accident and impelled toward eventual entropic dissolution. Within the bounds of such a philosophy, free will, purpose, and rational thought itself are drained of any significant meaning and rendered absurd, for of what possible value are reason and action if they are merely the necessary consequents of implacable natural laws?

Our common sense, our common rationality, comprises the elemental sense of mental proposition that invites and enables us to distinguish, assert, believe, and think anything at all. To deny that it exists is to contradict oneself: the "validity of rational thought," wrote Lewis, " . . . is the necessary presupposition of all other theorizing."[35] By implicitly undercutting the validity of rational thought, the modern naturalists render meaningless and self-contradictory all that they have to say. And, equally to the point, they fly in the face of common sense, of which the proposition that "neither Will nor Reason is the product of Nature" is a basic article, as Lewis quite rightly pointed out.[36] In practice no one assumes that stone or cactus or camel has any part of will or reason: man is a thing apart from the natural as well as a part of it. Assertions to the contrary constitute transgressions against common sense and elemental sanity; they are the legacy of nineteenth-century mechanistic scientism at its intellectually feeblest—a thesis for which Lewis never ceased to gather evidence and which he never ceased to argue.

To the "learned foolishness" of mechanistic scientism—by no means universal among scientists—Lewis's reply might have been like that of Blake, who in a different way and with mainly visionary rather than philosophical resources tried to fight the same "mind-forg'd manacles" in his own time: "trouble me not with thy self-righteousness./I have innocence to defend and ignorance to instruct."[37] Incomplete though it sometimes is, it is the essential sanity of innocence, of common sense, and of that "broad daylight of proportion" that Chesterton and Lewis always sought to defend.

12

This tradition of Godly common sense was itself the matrix out of which came the great developments of Western science, as the distinguished historian of science Stanley L. Jaki argued in his 1977 Fremantle Lectures at Oxford. "The new organon of science," he writes,

> was not in the voluminous fumbling of Bacon with mostly irrelevant facts but in the conviction shared long before him of the fact that since the world was rational it could be comprehended by the human mind, but as the product of the Creator it could not be derived from the mind of man, a creature.[38]

This idea of the rational intelligibility of the cosmos is one of the deepest and most pervasive intellectual assumptions of Western culture, although it may now be breaking down after three centuries of metaphysical imperialism on the part of what Whitehead deplored as "scientific materialism," which has left nature—and, we might add, man—in tatters.[39]

With his education in metaphysics, Lewis was well-prepared to criticize this modern scientism in detail, and it was one of the chief concerns of his life to do so; he was able to argue carefully and consistently the validity of rational thought as the premise of specifically human existence. "Unless all that we take to be knowledge is illusion," he once wrote, "we must hold that in thinking we are not reading rationality into an irrational universe but responding to a rationality with which the universe has always been saturated."[40] But one need not be a philosopher to realize this in a rudimentary form—it is an obvious dictate of common sense, fortified by the tradition of most wise men in the West, the tradition that Lewis communicates and extends. It is part of that legacy of the best that has been thought and said with which humane literacy can put us in touch.

A critic has written of the work of Norman Mailer what could be said of much in modern literature and thought generally, that it contains a "perverse exaltation of the submerged at the expense of the humanly self-evident."[41] By contrast, we could with some accuracy say of Lewis what Joseph Sobran has said of Samuel Johnson, that he "may fairly be called the great champion of the obvious."[42] On the appearance of Lewis's third series of BBC broadcast talks in book form in 1943, Robert Speaight wrote of him as broadcaster and writer that he "approaches you directly, as a rational person only to be persuaded by reason. He is confident and yet humble in his possession and propagation of truth."[43]

Having been badly wounded and finally invalided home during the First World War, Lewis later wrote that in the trenches "I came to know and pity and reverence the ordinary man."[44] In the same vein, writing to a friend in 1930 about the experience of reading George MacDonald's *Diary of an Old Soul*, Lewis spoke of his joy at finding himself "on the main road with all humanity."[45] Bede Griffiths wrote of Lewis in his youth that he "always affected . . . to be a plain, honest man" and that it was, "no doubt, the expression of a determination to avoid all pretentiousness, which later developed into a real and profound humility, but it . . . often concealed his real greatness."[46] And Griffiths, who had been Lewis's student, writes that he "always treated me as an equal in every respect, as I believe that he treated all his other friends. . . . I think that it was through him that I really discovered the meaning of friendship,"[47] a verdict borne out by Owen Barfield, among others, as well as by numerous strangers or correspondents who made visits to Lewis at Oxford or Cambridge.

Griffiths marvelled at the psychological insight of *The Screwtape Letters* and *Mere Christianity*, at Lewis's "capacity to speak to the common man and see into the hidden motives in the heart of everyman."[48] In his concern with everyman, Lewis was a democrat but never out of sentimental idealism or egalitarianism ("Equality," he said, "is a quantitative term and therefore love often knows nothing of it"[49]); like Swift he disliked and dreaded mankind in the mass and "mass man," and could only conceive of the willed loving of individuals which is Christian loving-kindness. He considered sin—excessive love of self and the resulting abuse of others—a fact of human nature and history; in the spirit of Augustine and Acton and Butterfield, of Montesquieu and the American founding fathers, he wrote in 1943 that the "real reason for democracy" is that man "is so fallen that no man can be trusted with unchecked power over his fellows."[50]

Lewis's attitudes toward the common man constituted yet another sense in which he stood in the tradition of Johnson and Chesterton, of whom John Gross has written that "[m]any other

English writers have preached Democracy in the abstract: Chesterton is one of the very few who genuinely liked the common man (because he was sure there was no such thing)."[51] Lewis too thought "there was no such thing," and knew that what is common to us all as human beings is not at all common. In his best-known sermon, "The Weight of Glory," preached in June of 1941 in St. Mary the Virgin Church in Oxford, Lewis said,

> There are no *ordinary* people. You have never talked to a mere mortal. Nations, cultures, arts, civilizations—these are mortal. . . . But it is immortals whom we joke with, work with, marry, snub, and exploit. . . . Next to the Blessed Sacrament itself, your neighbour is the holiest object presented to your senses.[52]

This is the doctrine of man as a holy object, *res sacra homo*, in which Lewis invested an interest as profound as his belief in the source from which it is derived—the God in whose image we are created. Lewis's demystifying vernacular, his direct appeal to the reason and dignity of the common man, and his unfailing personal humility and attendant reverence for the *imago Dei* he perceived in all men qualified him uniquely to make the case against modern scientism, to iterate eloquently the fact that man is infinitely more than merely a "part of nature, in the same sense that a stone is, or a cactus, or a camel."

II

Scientism vs. Sapientia

Reflecting on . . . being in the grip of his futile chase of Moby Dick . . . (Captain Ahab) could merely mutter, 'All my means are sane, my motives and objects are mad', and go on with his maddening endeavor. In an age of hydrogen bombs, multiple warheads, cruise missiles, laser guns, anti-satellite satellites, it is not even possible to say that all means are sane.

S. L. Jaki

*L*EWIS saw that the greatest attack in modern times on the common sense of *recta ratio* was coming from people, usually not scientists, whose philosophical beliefs and practical attitudes had nevertheless been mistakenly derived from the popularity, apparent simplicity, and evident power of the natural sciences. He was a partisan in a debate that has raged for centuries between those who assert the primacy of metaphysical knowledge and those who argue for the priority of physical reality. This is an old and important conflict, one fought with asperity in the seventeenth century between men such as Donne, Milton, and the Cambridge Platonists on one side, and Bacon and the Royal Society on the other—the conflict between the Ancients and the Moderns, with philosophers, poets, and theologians ranged against some scientists and many more enthusiasts of science.

In the eighteenth century, men such as Bishops Butler and Berkeley and the Tory satirists Swift, Pope, and Johnson opposed the naturalism of Hume and Bentham and the French philosophes. All the while, the same conflict was raging across Europe, and by the nineteenth century it pervaded the whole of Western culture. In our time it has often been posed in terms of a schism between religion and the humanities on the one hand and science and technology on the other, and recently and specifically it emerged in the debate between C. P. Snow and F. R. Leavis about which of the "two cultures" should have priority in education and life.

A somewhat more detailed history of this conflict and a consideration of its issues are necessary in order to understand Lewis's role in it and its relevance to us now, but it is important to bear in mind at the outset that the battle lines shift considerably and that there are many figures in the history of European thought whose role in these disputes is the opposite of what would have

18

been expected. Bacon, for instance, although a propagandist of science, was not a scientist at all, and in him was lodged much of the older philosophical tradition; and Copernicus and Newton were devout Christians as well as great scientists. Generalizations at this level about the history of ideas are often so broad as to be nearly useless. The detailed modern studies of this terrain, such as the "background books" of Basil Willey, Alfred North Whitehead's *Science and the Modern World*, Herbert Butterfield's *Origins of Modern Science*, Arthur Koestler's *The Sleepwalkers*, and the recent work of S. L. Jaki, give much of the grain and fiber of the developments since the seventeenth century. They are indispensable for an understanding of the genesis and development of beliefs that all of us hold both implicitly and explicitly. As Dr. Johnson said, most of us "catch our opinions by contagion," rather than by calculation, from the unremarkable influences of tradition and our culture's prevailing climate of opinion—in some ways no bad thing in itself, if we are aware of the process.

Our consideration can begin with the justly celebrated remark made at the beginning of the modern era by that wise jester François Rabelais, who started out his life as a Franciscan but before he had finished it is said to have experimented on corpses: *"La science sans conscience, ce n'est rien que la mort de l'âme"*[1]—science without conscience is nothing but the death of the spirit—an insight encompassing much of the triumph and tragedy of the era since the Renaissance and containing the kernel of Lewis's case against modern scientism. The adage has an interesting, indeed a haunting, modern echo in the remark attributed to Einstein, and characteristic of his cosmic piety, that "religion without science is lame, but science without religion is blind."

In a recent retrospective review of the life work of the French philosopher Jacques Maritain can be found a word that impinges yet more sharply upon this issue: *sapienza*—a word which, still vital in the Italian, is a closer linguistic heir of the Greco-Latinate classical and Christian tradition than any equivalent term in French or English could lay claim to being:

> Maritain sa anche che la scienza senza la sapienza è cieca, e aggiunge che la salvezza della cultura umana nella sua totalità

19

> consiste nel non perdere mai di vista l'importanza della sapienza metafisica.[2]

> Maritain knows also that science without *sapientia* [sapience, wisdom] is blind, and he adds that the salvation of human culture in its totality consists in never losing sight of the importance of metaphysical *sapientia*.

Metaphysical wisdom—*sapientia*—is a specific mode of knowing that Lewis sought to defend and propagate in his writing; it is at the core of the perennial philosophy that has seemed to grow so weak in the modern era, as *homo sapiens* has waned and *homo sciens* has waxed.

Issues such as the procedures and validity of rational thought and argument are presuppositions on which scientific thought and experiment rest, but they are themselves not "scientific": they are philosophical. Science depends on philosophy for the validity of its terms and procedures and the determination of the uses to which scientific knowledge will be put. To say, with the radical empiricist, that only factual statements have validity is to be not only dogmatic but self-contradictory, since the statement itself is not factual. Or as Lord Hailsham, who had a philosophical education at Oxford similar to Lewis's, recently put it in his autobiography,

> while . . . it may be true that [within] the field of science, its terms of reference . . . , are circumscribed by the proposition that only that exists which can be measured or observed, the proposition itself is one which cannot itself be observed, and is therefore one which cannot be true of all being and if it is asserted as such becomes immediately self-contradictory.[3]

It is thus a misuse of science, an instance of scientism, the misapplication of scientific method, to assert that "what is not in principle observable is not in fact in existence." Truth, meaning, purpose, goodness, importance, are none of them scientific facts: they are, as Lewis says, "wholly immaterial relations." And yet, the material world has always exercised the powerful appeal of an exclusive claim to reality—what Lewis himself calls "the sweet poison of the false infinite."[4]

Thought is itself infinite, defying adequate description or

definition in merely physical categories or terms, as the noted brain surgeon and neurologist Wilder Penfield recently concluded after a lifetime of thought and research. Throughout his career, Penfield confesses, he had, like many scientists, tried to prove that "brain mechanisms account for the mind," but he finally came to the conclusion that "because it seems to me that it will always be quite impossible to explain the mind on the basis of neuronal action within the brain, and because it seems to me that the mind develops and matures independently throughout an individual's life as though it were a continuing element . . . I am forced to choose the proposition that our being is to be explained on the basis of two fundamental elements," material and immaterial, physical and metaphysical.[5] It is this moderate dualism that constitutes an important part of the "perennial philosophy," the *consensus gentium*, the common sense of mankind. Normally, as Professor Owen Chadwick says, "The common sense of the human race finds it as obvious that everything is not matter as it is obvious that everything is not in the mind."[6]

It has long been popular to deny this obvious dualism, as most modern naturalists implicitly or explicitly do, caught up as they are in the legacy of nineteenth-century scientistic materialism, and especially of T. H. Huxley's "epiphenomenalism." But such an assertion necessarily negates the validity of all statements, including the assertion itself, as Lewis obstinately and unfashionably insisted; he agreed with Whitehead that scientists "animated by the purpose of proving that they are purposeless constitute an interesting subject for study."[7] S. L. Jaki, too, has noted how very odd it was that "science, which is a most purposeful endeavour aimed at the understanding and control of the physical, has been unable to foster a vivid appreciation in man for an understanding of purpose."[8] But perhaps the only duty of science in this regard is to confess its inadequacy in the matter. The limits of its method help to illuminate and point to the philosophical realm for the understanding and appreciation of realities such as purpose.

In pursuit of the historical genesis of this scientistic climate of opinion, we would do well to consider a comment made by Leibniz in 1687 (the year that Newton published his *Principia*), that mechanistic science seemed to him a support for atheists because consideration of purpose was excluded by the very means in which it was conducted.[9] This tendency had been implicit in much of Bacon's propagandistic writing for the "new philosophy" of the seventeenth century, and it has more and more explicitly characterized the utilitarianism which derives from Bacon. *"Nam et ipsa scientia potestas est"*—scientific knowledge is power, Bacon optimistically wrote. His famous words have been rendered cruelly, ominously ironic with the advent of megadeath weaponry, stockpiled pathogens, experimentation on live fetuses, neutron bombs, nerve gas, and laser guns.

The true object of scientism, wrote Lewis quoting an equally famous assertion of Bacon, "is to extend Man's power to the performance of all things possible."[10] But "all things possible," as we have seen in our century, is a menacingly amoral category; as one of the scientists who worked on the atomic bomb later said regretfully, "We felt that neither the good nor the evil applications [of the bomb] were our responsibility." Such irresponsible realization of all things potential regardless of consequences is a catastrophic result of the rational and moral bankruptcy of scientism.

The great English philosophical poets have always seemed to understand that although the case against scientism is implicitly argued by Christianity in word and sacrament, and is conclusive to the educated rational man and theoretically available to all persons, some more popular form than rational argument is nevertheless necessary to counteract its more immediately tangible and visible appeals. The positive tactic that the great metaphysical poets employed was to affirm the reality of the two worlds, the physical *and* the metaphysical, which *homo sapiens* simultaneously inhabits. Their work serves to draw the reader up the Platonic ladder of ascent (and assent) in the epic poetry of edification, allowing him in the process of reading to experience, to taste the intelligible Good by transcending the physical and experiencing a conscious participation and exaltation in the metaphysical realm. This is the genius of Dante, of Spenser, and of Milton; in a more popular key it is also characteristic of Langland and Bunyan, and, in lyric form, in the aptly named Metaphysical Poets of the Renaissance and seventeenth century. Donne invites the ascent from the experience of spatial and temporal sensations to the transcendent rationality that orders and regulates them:

22

When wilt thou shake off this pedantry
Of being taught by sense, and fantasy?
Thou look'st through spectacles; small things seem great
Below; but up into the watch-tower get,
And see all things despoiled of fallacies.[12]

In addition to the poetry of affirmation, there grew up a poetry and prose of satire that attempted to reveal the inherently contradictory nature of scientistic naturalism by ridiculing it through an implicit appeal to the reader's common sense. With its lack of a transcendent vantage point from which humanly to estimate, evaluate, and regulate nature, scientism flounders between the paradoxical extremes of glorification and debasement in its approach to nature. "Man can only gain control over nature by obeying it,"[13] wrote Bacon; "we must put nature to the rack, to compel it to answer our questions."[14] And his intellectual descendant Diderot later wrote that man was learning to "torment nature" into giving up her secrets for the increase of man's power and pleasure. The ominous and even tragic overtones of such statements made scientism an inviting target for satirical dissection.

The Tory satirists Swift, Pope, and Johnson, of whom Lewis is an intellectual heir, devoted much of their writing to an incisive criticism of naturalism, often trying explicitly to put the philosophical objections to naturalism in a rhythmic and popular form, as in these lines from the *Dunciad*:

> *Philosophy*, that leaned on Heaven before,
> Shrinks to her second cause, and is no more.
> [Bk. IV, II 643-44]

With his own unique wry elegance, Pope gives voice to an attitude reasonably widespread among his intellectual compatriots concerning the dire portents of the growth of scientism. From the outset it was clear to thoughtful analysts that natural philosophy ungrounded in the metaphysical realities had the strength to crush philosophy, impugn religion, and subvert ethics among its unthinking disciples.

> *Physic* of *Metaphysic* begs defense,
> And *Metaphysic* calls for aid on *Sense*!

23

. . . .
Religion blushing veils her sacred fires,
And unawares *Morality* expires.
[Bk. IV, II.645-50]

In *Gulliver's Travels*, Swift's satire on the excessive claims made for experimentation and technological innovation takes the form of the "Academy of Projectors of Lagado," one of whose projects is to replace words with things so as to be closer to "empirical reality." Everyone is required to carry a pack of objects around on his back for use as devices of communication, since words had been banned as misleading. The opposition to this project comes from "the women in conjunction with the vulgar and illiterate," who threaten to rebel "unless they might be allowed the liberty to speak with their tongues." Swift's satirical remark is relevant to our theme: "such constant irreconcileable enemies to science are the common people."[15]

An historian of science describes the prophetic quality of Swift's fable in a comment also relevant to the passage from the *Dunciad* when he notes that a few years later Hume was to recommend "the committing to flames of all books that contained no quantities and matters of fact":

> Such was in Hume's radical empiricism or sheer sensationism the beginning of the ultimate elimination of the common sense of all people, common or not, through the discarding of many a word in favour of sensous objects, or rather of mere sensations, a programme akin to the one advocated in the Academy of Projectors of Lagado.[16]

Pope and Swift were thus not shadowboxing, but instead, on the basis of a profound intuition, protesting and satirizing the tyranny of scientistic thinking which attributes validity (itself not a quantity) only to quantity, only to the merely physical and material. They attempt to revive, nourish, and protect the common reason of man against specialists who would reduce it to sense perception only. Their more or less intuitive satire set the stage for Lewis's hero Samuel Johnson.

All of Johnson's conversation, poetry, and prose are didactic and religious, embodying an accumulation of experience, thought, and reading, an incarnation of tradition at its most vital and lucid, that make him one of the great sages of the world, a priest of *sapientia*, the common man's intelligible and trustworthy philosopher. In his biography of Milton, Johnson warned against the encroaching naturalism and the restless spirit of eighteenth-century innovation and promoted the elementary metaphysical and moral truths of *sapientia* without which science itself is not only self-contradictory but, to use Einstein's term, "blind." As he characteristically notes,

> The truth is that the knowledge of external nature, and the sciences which that knowledge requires or includes, are not the great or the frequent business of the human mind. Whether we provide for action or conversation, whether we wish to be useful or pleasing, the first requisite is the religious and moral knowledge of right and wrong: the next is an acquaintance with the history of mankind, and with those examples which may be said to embody truth, and prove by events the reasonableness of opinions.

"Prudence and justice," Johnson points out, "are virtues and excellences of all times and all places," whereas specialized knowledge of the sciences cannot be either necessary or normative for life and must remain subordinate to the moral rationality which it is every person's obligation to acquire and apply. As he puts it most crucially and convincingly,

> we are perpetually moralists, but we are geometricians only by chance. Our intercourse with intellectual nature is necessary; our speculations upon matter are voluntary and at leisure.

And again, scientific expertise, which he calls "physiological learning,"

> is of such rare emergence that one man may know another half his life without being able to estimate his skill in hydrostatics or astronomy; but his moral and prudential character immediately appears.

To unseat *sapientia* and seek first scientific knowledge, Johnson would have us know, is a grievous misapplication of priorities; not only does such an error place the cart before the horse—it slays the horse altogether.

By pressing his criticism of the excessive claims made for *scientia*—empirical knowledge of the natural world—and by serving as a defender of the vital necessity and priority of *sapientia* in any reasonable scientific endeavor, Johnson took his place as a figure within a centuries-old tradition sustained by a host of other important figures. Among his significant predecessors was Socrates, whose labor Johnson said it had been

> to turn philosophy from the study of nature to speculations upon life; but the innovators whom I oppose are turning off attention from life to nature. They seem to think that we are placed here to watch the growth of plants, or the motions of the stars. Socrates was rather of opinion that what we had to learn was, how to do good, and avoid evil.[17]

Johnson followed Socrates in recognizing that *sapientia* comprises a genuine rational knowledge of one's own nature *and* of transcendent metaphysical realities. Cicero had defined *sapientia* as the knowledge of "the bonds of union between divinity and man and the relations of man to man"[18]—knowledge which *homo sapiens* is uniquely fitted and obliged to possess and which regulates all our knowledge of physical nature. Aquinas was later to say that at its highest this knowledge brings its possessor into intimate contact with God Himself: *"Deus est ipsa sapientia."* It is from within this Christian humanist tradition that Johnson speaks, in a kind of apostolic succession of humbly but doggedly virtuous men who see virtue as man's "natural" goal and rule, but who also resist the pharisaical temptation of self-satisfied pride in virtue that characterized many classical moralists, and especially some Stoics. Virtue is inferred, recognized, and venerated according to a divine pattern to which we owe gratitude and piety. "Human nature is not sufficient unto itself," Johnson wrote vehemently, "as the over-proud Stoic sect deceitfully boasts."[19]

Outstanding passages from the recorded heritage of the Christian humanist tradition, such as Johnson's tribute to Socrates, served as touchstones for Lewis in his project of defending and applying the common reason of man in addition to providing models for his own methods and arguments, especially for his own great short philosophical work *The Abolition of Man*. It is not the specialist's expertise, not the scientific knowledge of many things that is needful to everyman; instead, what is called for is the reception, possession, and practice of rationality in its elementary moral and philosophical forms, what Johnson calls "our intercourse with intellectual nature." This intercourse provides the basis for both validity and morality, and it makes possible the specifically public ethical order that forms the *res publica* without which society and culture cannot exist, without which man becomes a barbarian.

In the late eighteenth century, however, the triumphs of technical acumen, especially in invention and industry, increased the lure of the knowledge of things, despite the calamitous fact of the French Revolution. The mechanistic, contradictory, restless materialism of the French philosophes, Chateaubriand wrote, "is already forgotten, and all that remains is the French Revolution";[20] but their scientific materialism was not really forgotten. Its impact encouraged the "performance of all things possible," and especially the development of the factory system, generated to a great extent by a pervasive fascination with material power, an increasing perfection of technical means ungoverned by any sure knowledge of the ends to which they should be directed.

Of the causes of the French Revolution Arnold Toynbee wrote toward the end of his life that "a sinister ancient religion which had been dormant suddenly re-erupted with elemental violence." The religion to which he referred was that most deadly phenomenon, "the fanatical worship of collective human power." The appalling Reign of Terror, he went on, "was only the first of the mass-crimes that have been committed during the last hundred and seventy years in this evil religion's name."[21] It is this religion and this historical concentration of amoral knowledge and power of the few over the many and over nature itself that Lewis called "that hideous strength."

The radical subjectivism of men such as Nietzsche, who proposed that life is an aesthetic phenomenon and that man is "beyond good and evil," and the radical scientism (objectivism) of men such as B. F. Skinner, who propose that man is beyond freedom and dignity, both violate common sense and lead to amorality and, logically, to barbarism. The intellectual forebears of both sides of this dissociation of sensibility can be found in the eighteenth century: the sensationalist Hume, and more explicitly

Rousseau and Sade, are the antecedents of such romantic primitivists as Nietzsche and our hippies and Hell's Angels: Diderot (who would "torment nature" to help increase our power and pleasure), D'Holbach, LaMettrie, and Bentham are the forefathers of our Marxists and our behaviorists. Most men of the nineteenth century knew nothing, of course, of the direction of these currents of thought as the appeal of scientism gnawed more and more deeply into the minds of the European intelligentsia and middle classes. The dissenters who saw and fought scientism were usually imaginative writers, theologians, or philosophers, but it is well to bear in mind that many scientific men themselves were also innocent of participation in the religion of science propagated by their over-enthusiastic and often unwelcome admirers.

The intoxicating vision of *scientia potestas est*, Russell's "power knowledge" unregulated by conscience or *sapientia*, produced what often seemed miraculous transformations of landscape and living conditions in nineteenth-century Europe, transformations that nowadays seem in large part to be quite clearly regrettable. Certainly they were resented by the victims upon whom they were thrust, a fact attested to by the great history and literature of social protest, including Goldsmith's "Deserted Village," the Luddites, Blake's prophecies, Dickens's novels, Ruskin's essays on "illth," and the growth of Marxism and trade unionism. In many ways Blake's criticism of industrialism and amoral rationality—what sociologists nowadays call "functional rationality"—was the most profound, but it was unavailable to his rationalistic contemporaries. His Christian and Platonist hostility to radical empiricism and functional rationality is of the depth, intensity, and ethical insight of the Old Testament prophets, but like many of them he too, in his own words, "sold wisdom in the marketplace where none came to buy."

Much more of the Romantic reaction to the eighteenth-century rationalistic scientism consisted of a complex nostalgia for Nature and the flight from a world stripped bare and corroded by analysis, "sicklied o'er with the pale cast of thought." According to Lionel Trilling, "Wordsworth's great autobiographical poem *The Prelude* gives the classic account of the damage done to the mind of the individual, to its powers of cognition no less than to its vital force, by the scientistic conception of mind that prevailed among intellectuals at the time of the French Revolution."[22]

There were some Romantic dissenters who criticized scientism with some effect in the short term, sometimes consistently and sometimes inconsistently, in intuitive flashes of philosophical or religious insight. Goethe was a divided soul, his Faust an exemplar of the sickness of the restless lust for manipulative knowledge (*libido sciendi*) at the expense of philosophical and moral knowledge (*philo sophia*) who was nevertheless sometimes aware of the nature of his own disease:

> And here I stand with all my lore
> No wiser than I was before.

But, asks Erich Heller, "What is Faust's sin? The restlessness of spirit. What is Faust's salvation? The restlessness of spirit."[23] Striving is his salvation, in the mode of modern busy-ness:

> Whoever strives with all his power,
> We are allowed to save.

This is the melodramatic romantic self, romantic Prometheanism and Titanism, heroic vitalism, part of the problem rather than the solution. Yet Goethe, an immense genius, could also pour the wine of *sapientia*, as when he wrote that "everything that liberates our minds without at the same time adding to our resources of self-mastery is pernicious."[24] Here the moral realm is touched and illuminated.

Some of the *sapientia* of Johnson and Burke, and of the German critics of French mechanistic thinking, was recovered and transmitted by Coleridge, whose agonized soul and mind were in full flight from British utilitarianism. He was appalled by the narrowness of strict empiricism: "are not the experimentalists credulous even to madness," he asked, "in believing any absurdity, rather than believe the grandest truths, if they have not the testimony of their own senses in their favor?"[25] He ridiculed as well those who abandoned as meaningless such perennial moral issues as the problem of evil on the grounds that such matters do not admit of experimental investigation and proof.

Coleridge perceived that in a significant sense, those who profess an interest solely in what is empirically verifiable abdicate their responsibility as *homo sapiens*; their hunger for the immediate, for the perpetually new and topical, cuts them off from the sort of speculation and meditation that might yield real knowledge about their inner selves and the metaphysical realities underpinning the natural world. The moderns want, Coleridge said, "something new, something *out* of themselves—for whatever is *in* them, is deep within them, must be as *old as* elementary Nature."[26] In Plato, Shakespeare, the seventeenth-century divines and poets, and in the Bible, we come into direct and personal confrontation with this "elementary Nature," and yet, according to Coleridge, the Bible has become almost impossible to *hear* and take seriously because of "the notion that you are already acquainted with its contents." As scientistic, quantitative thinking increases among individuals and in society, *sapientia* declines, and truth loses its lustre. "Truths of all others the most awful and mysterious," Coleridge writes, come to be ignored or to be "considered as so true as to lose all the powers of truth, and lie bedridden in the dormitory of the soul."[27] The Bible, "the foundation-stone of Western civilization," Basil Willey writes, is for Coleridge "the most potent symbol of that ancient wisdom from which the modern world was in apostasy."[28]

It was, however, Kierkegaard who was among the most systematically clear-sighted enemies of the belief in scientific materialism in the nineteenth century. His assertion that "[t]he bourgeois mind is really the inability to rise above the absolute reality of time and space," is as precise a short definition of scientism as one could want.[29] Like Johnson, he invoked the authority of Socrates in his struggle against the sophistical naturalists, and like Swift he employed cutting satire. "If the natural sciences had been developed in Socrates' day as they are now," he writes in *The Present Age*,

> all the sophists would have been scientists. One would have hung a microscope outside his shop in order to attract custom, and then would have had a sign painted saying: "Learn and see through a giant microscope how a man thinks" (and on reading the advertisement Socrates would have said: "that is how men who do not think behave").[30]

The specialization and logic-chopping brought about by the onslaught of mere "functional rationality," devoid of ethics and common sense, Kierkegaard scornfully called "the increasing mass of drivel which is called science," and he wrote ominously of the devaluing process brought about by increasing numbers of people who believe only in quantity and technique without referring them to any general value or subordinating them to any moral norm or end: "In the end, all corruption will come from the natural sciences."[31] Here, hauntingly, we have the precise echo of Rabelais's "science without conscience" presiding over "the death of the soul," the extinction of the specifically human and moral aspect of the mind.

Although Kierkegaard tended to a radically fideistic Christianity that often mistrusted reason's claims, his hostility was not to science itself but to excessive and misapplied extrapolations from it, to the obsession with means at the expense of ends, with technique at the expense of purpose. The restless Faustian desire for knowledge (which the Medievals called the *libido sciendi*) as an instrument with which egotistically and amorally to exploit nature and other men (which the Medievals called the *libido dominandi*) Kierkegaard profoundly mistrusted: "Almost everything that flourishes nowadays under the name of science . . . is not science at all but curiosity."[32] As John Passmore has written, Kierkegaard "was happy to let science deal with plants and animals and stars"; but, Kierkegaard wrote, "to handle the spirit of man in such a fashion is blasphemy."[33]

The case against considering man a thing, a "common object of the seashore," lies at the heart of the critique of scientism as Lewis makes it and as Martin Buber proposed it in his distinction between "I-it" and "I-Thou" relationships. The distinction between means and ends, between things and objects on the one hand and persons and essences on the other is at the root of moral culture and civilization; it is a distinction that the characteristic procedures and terms of the natural sciences can neither discern nor make without violence, contradiction, and confusion, and for which they must therefore depend on philosophy and religion.

With his wayward and unsystematic genius, Nietzsche too had profound insights concerning the roots of scientism. In *The Gay Science* he questioned whether the "ultimate goal of science is to create for man the greatest possible amount of pleasure and the least possible amount of pain,"[34] implicitly underscoring the hedonism of scientific materialism, devoid of and destructive of ethics and common purpose. If the maximization of pleasure and the minimization of pain is the only norm of obligation that need be attended to, then how one defines and pursues pleasure is his own business, and no *general* obligation to work for a theoretical "greatest good of the greatest number" can be pressed on anyone. Nor can a person be expected to pursue pleasure only so long as it doesn't hurt others—all grounds for condemning even destructive behavior have themselves been destroyed with the initial assertion that pleasure is the highest good. All moral assertions are relativized and destroyed and life becomes, as Nietzsche said, an aesthetic phenomenon. It was the realization of this fact that led Lewis to criticize utilitarianism as "futilitarianism," because the working out of its principles in society has inevitably led either to hopeless muddle or to calamity. In the political world, for instance, utilitarianism has led either to what F. R. Leavis aptly called "Techno-Benthamism" or to the more spectacularly disastrous Marxist scientism—"scientific socialism"—with its empty promise of historical inevitability and its tragically real tyranny and barbarism.

The objections to the tide of scientism were not, of course, felt solely in Europe; in America too the cry went up, and it was often the imaginative writer who got an intuitive grasp of its drift and its demoralizing effects. Melville wrote despairingly of

> Man disennobled—brutalized
> By popular science—atheized
> Into a smatterer—[35]

And Emerson, who was confused and confusing about much else, nevertheless penned a small poem in which he provided a splendid illumination of what happens when the right relation between metaphysical and physical modes and kinds of knowledge is not preserved—when *scientia* is allowed to destroy *sapientia*:

> There are two laws discrete,
> Not reconciled,—
> Law for man, and law for thing;
> The last builds town and fleet,
> But it runs wild,
> And doth the man unking.[36]

32

"Things are in the saddle," he also wrote, "and ride mankind."
And again, "A terrible machine has possessed itself of the ground,
the air, and the men and women, and hardly even thought is
free."[37] Second things suffer, as Lewis was fond of repeating, when
put first. Science is a good servant but a bad master, a good
method for investigating and manipulating the material world, but
no method at all for deciding what to do with the knowledge and
power acquired thereby.

Nevertheless, with the industrial miracle (which was for so
many people a nightmare) and the resulting popularization of the
power of scientific methods, the appeal of scientific materialism,
especially to the middle classes, increased in the nineteenth century, especially after the Darwinian controversies. The German
materialist Karl Vogt lectured across Europe on Darwin and scientific materialism, propagating a harsh anti-religious and atheistic
philosophy. One of his most famous sayings was quoted over and
over again: "Thoughts come out of the brain as gall from the liver,
or urine from the kidneys." Feuerbach asserted that "Man is what
he eats," and commented dismally that "It used to be said 'In the
beginning God.' Now it is said, 'In the beginning the belly.' "[38] In
Disraeli's novel *Lothair*, the Monsignore identifies the elements
in what he calls the new "religion of science":

> Instead of Adam, our ancestry is traced to the most gro-
> tesque of creatures; thought is phosphorous, the soul com-
> plex nerves, and our moral sense a secretion of sugar.[39]

The chief popularizer of Darwin in England was Huxley, who
was as ethically earnest as he was rationally contradictory with his
theory of epiphenomenalism. He argued that consciousness, and
therefore thought, are merely by-products of the functioning of
brain chemistry and mechanisms; of course, such a view invali-
dates the validity and significance of thought itself, including,
necessarily, the thought that engendered Huxley's assertion of
epiphenomenalism. The forms of modern scientific determinism
that have come in Huxley's wake have often shared his self-
contradictions without imitating his strenuous ethical earnest-
ness, which was largely a result of the moralistic Victorian environment in which he was raised and in which he lived.

An earlier reaction against the scientism which Huxley pro-
pounded had come from Dr. Thomas Arnold, Headmaster of

Rugby School, who introduced educational reforms that were widely imitated at other British public schools and at American private schools, some of which were inspired and founded on his ideals. "Rather than have physical science the principal thing in my son's mind," he said, "I would gladly have him think that the sun went round the earth. . . . Surely the one thing needful for an Englishman to study is Christian moral and political philosophy."[40] In the 1881–82 debate between Arnold's son Matthew and T. H. Huxley, the battle was continued, but in nineteenth-century England the spirit of Johnson—and of Butler and Hooker and Erasmus before him—found vibrant new life in the writings of John Henry Newman.

The common-sense dualism that discerns and distinguishes mind and matter, God and the world, and attests to the reality of both, is frequent in Newman's thought: "that a divine influence moves the will," he wrote, "is a subject of thought not more mysterious than the result of volition on our muscles."[41] But in addition to theology and epistemology, Newman's great achievement and legacy lies in the writing on education, the *Idea of a University*. In that monumental work he employed the balanced language of moral reason in defense of the integrity and validity of the traditional modes of knowledge in the university and against the scientistic determinism that would destroy those modes of knowledge in the name of "reform" and "relevance": "I observe . . . that if you drop any science out of the circle of knowledge, you cannot keep its place vacant for it; that science is forgotten; the other sciences close up, or, in other words, they exceed their proper bounds, and intrude where they have no right."[42] Further, the "Religious Truth [of theology and metaphysics] is not only a portion, but a condition of general knowledge," without which it is impossible for us to have sound, consistent, and connected views. "To blot out [Religious Truth] is nothing short . . . of unravelling the web of University Teaching."[43] Writing in precise defense of *sapientia*, Newman says that to have "even a portion of this illuminative reason and true philosophy is the highest state to which nature can aspire, in the way of intellect."[44]

In Newman the voice of the great central tradition of metaphysical common sense speaks, the voice of Plato, of Aristotle, of St. John's Gospel, of Cicero and St. Augustine, of St. Thomas and Erasmus and Hooker and Johnson. It is a voice and an intelligence that discerns and defends the validity of the natural sciences, but which doggedly opposes the inappropriate application of their assumptions and methods and the excessive claims for their regulative status in life and curricula. These applications and claims constitute *scientism*, which Newman precisely defines as "an evident deflection or exorbitance of science from its proper course."[45] As an individual's common sense integrates and relates the findings of the mind and the separate senses, so "illuminative reason"—*sapientia*—regulates and synthesizes the methods and findings of the separate analytical sciences. *Sapientia* is an exalted form of common sense.

Reviewing a definitive edition of Newman's *Idea*, Henry Chadwick wrote,

> A characteristic feature of the modern university, so abhorrent to him, is precisely what he foresaw, namely, the pursuit of unrelated specialisms by individuals sharing few common convictions, and perhaps setting up as their ideal and criterion of all knowledge disciplines such as pure mathematics or physical science which appear furthest removed from moral values.[46]

The vast intellectual and moral chaos and despair of our time have been brought about largely by this transformation of the university into a cafeteria or smorgasbord, into a "multiversity." This fundamentally degenerative transformation is a chief characteristic and effect of the loss of faith in illuminative reason and of the failure of nerve, the abdication of moral authority by the professoriat, a feckless but continued "treason of the clerks" often of course popular with students in the short term, but disastrous to them in the long run. To the extent that order and purpose exist in the universities of the English-speaking world today, they are largely the legacy of Newman's eloquent philosophical traditionalism, the beacon light of which, clearly or dimly perceived, has illuminated and kept open the path to truth.

But the compulsive spirit of innovation, the lust for change and the new, which Arnold and Newman fought in related ways in the educational realm, was a chief effect of the intoxications of scientism, and it has continued to increase in effect in many other areas of modern life. "Each area of contemporary social life is impressed, not so much by the *content* of science," which in most cases is generally incomprehensible, "as by the *pace* of scientific discovery," notes sociologist David Martin; "the field of education feels the need to produce bogus innovation in order to show that it emulates the scientific paradigm; similarly so the church." And in this way, Martin concludes, "the notion of passing on a good from generation to generation is undermined," and the idea of the permanent validity of moral reason is weakened.[47]

Among the Victorians, Ruskin was one of the keenest critics of the encroaching scientific materialism but also one whose life presents the spectacle of an agonizing emotional and intellectual crucifixion, a madness partly due to the triumph over mind and landscape of the rapacious industrialism he so abhorred. He criticized "the thrill of scientific vanity" and the resulting sensationalism and neophilia of the restless nineteenth century and of much of its art, "the modern infidel imagination, amusing itself with destruction of the body, and busying itself with aberration of the mind." Criticizing the materialists, he sardonically remarked upon "the inconvenience of the reappearance of a soul among these corpuscular structures," and asked if "it were as just as it is convenient to consider men as actuated by no other moral influences than those which affect rats or swine." And in the moral sphere, it is not, he wrote, "by 'science' of any kind that men were ever intended to be set at one."[48]

The climate of opinion established by the materialistic eighteenth-century philosophes was extended by nineteenth-century positivists and evangelists of the idea of collective progress, such as Comte and Renan. Indeed, one historian of philosophy credits Renan with having given "birth to the first religion of science,"[49] and the cultural historian Edward Said writes that Renan throughout his career "seemed to imagine the role of science in human life as (and I [Said] quote in translation as literally as I can) 'telling (speaking or articulating) definitively to man the word (logos?) of things.'"[50] Said precisely describes here the confusion of categories characteristic of scientism: it mistakes the truth about quantities, material and spatial realities, for the Logos, the Word of *sapientia*, the realm of qualities, purposes, values, ends.

But for all the aesthetic and materialistic confusion of the French tradition, it had had in the seventeenth century its own standing reproach and cure in the work of Blaise Pascal, himself a great scientist as well as philosopher. The scientistic eighteenth-century philosophes had denied or attacked his work, but it remained alive and was later to be recalled by real scientists such as the profound twentieth-century physicist, philosopher, and historian of science Pierre Duhem, who argued for the autonomy of metaphysics and religion and their priority over science. Duhem knew Pascal's *Pensées* by heart.

Earlier in the nineteenth century, Kierkegaard's scorn for scientism was equalled perhaps only by that of Dostoevsky. Prophet that he was, he seems to have had a bitter foretaste of the brutally malignant effects that would derive from the revolutionary child of utilitarianism, that contradictory materialistic-moralistic totem Marxism, which called itself— characteristically—"scientific socialism." Nothing could be more obviously and disastrously scientistic than the ideology of scientific socialism. It was to become within fifty years of Dostoevsky's death the most tyrannical and morbid religion in history, a crematorium for life and values, and with Nazism, the conclusive example of Toynbee's contention about the barbaric worship of collective human power that had begun to replace Christianity in the eighteenth century.

Dostoevsky's most systematically venomous mockery of nineteenth-century scientific materialism is, of course, to be found in *Notes from Underground*, but can be found in his other works too, such as *The Possessed*. "This has been the special knack of pseudo-science," he writes,

> that terrible scourge of mankind, a scourge worse than plague, famine and war, an evil that didn't exist until this century. Half-knowledge is a tyrant without precedent, one that has its own priests and slaves; a tyrant that is worshipped with unprecedented awe and adulation and before which science itself fawns and cringes.[51]

Dostoevsky prophetically foresaw the "treason of the clerks" that would be committed by so many intellectuals, but especially "scientific" ones, in their enthusiasm and support for "the Soviet experiment" from the 1920s until quite recently. As Orwell was to write in 1946 about such supporters of the Soviet Union as the Cambridge scientist J. D. Bernal, "They appear to think that the destruction of liberty"—and, we might add, of life—"is of no importance so long as their own line of work is for the moment unaffected."[52] It is revealing, too, that in addition to the Gulag, today "physical science . . . flourishes in a Soviet Union in which creative literature survives only as an underground activity, and art is undisguised propaganda."[53] Of this scourge, Dostoevsky was the extraordinarily prescient and discomfitting prophet, and Alexander Solzhenitsyn is his heir.

But there was another writer who added his distinctive and wonderfully apt insights to the scientism controversy, a man whose life straddled the nineteenth and twentieth centuries and whose temperament and style were altogether different from those of Pascal, Kierkegaard, and Dostoevsky. He was altogether as cogent and convincing in his arguments as were his compatriots in the attack upon scientism, and yet he presented his ideas in an entertainingly light-hearted manner. If he does not deserve the title of prophet, perhaps he can best be identified as a holy fool, wiser than the worldly wise or prudent; his jesting was the vehicle of a profound wisdom that might well be described as grave levity. G. K. Chesterton was a gregarious, Rabelaisian figure of immense intellectual generosity, a man so elementally sane and joyous that he almost seems out of place in the twentieth century. Although he too wrestled with the nightmare and despair that so deeply scarred Dostoevsky, he kept it very much to himself, for there is little evidence of it in his work—which remains, however, at the furthest point imaginable from superficial optimism. He is the inheritor and spokesman of the exalted common sense of St. Thomas and Erasmus and Hooker, of Johnson and Burke and Newman, and no one in our century has had and shared with the world more wisdom, goodness, and humorous but profound loving-kindness. As Dr. Johnson's generosity, eloquence, and dogged common sense were a model and inspiration for Chesterton, so were those qualities in Chesterton an inspiration to C. S. Lewis.

Chesterton was, in the words of the great twentieth-century

French Thomist philosopher and historian Etienne Gilson, "one of the deepest thinkers who ever existed," despite the fact that he was neither a scholar, nor a systematic philosopher, nor even what we now call an "intellectual." He was instead like the early Swift and Johnson, a journalist, and even, in a sense, a hack writer; but he was a metaphysical journalist, a hack writer who "saw life steadily and saw it whole." In the tradition of Arnold, he tried to get what he believed was every person's right and duty to get, what he called the "vision of life in the light of a general philosophy." In an age of specialists, he was a generalist; in an age of philosophy professors, he was a philosopher; in an age of ephemeral news, he strove, as an imitative reviewer said in *The Listener* in 1942, to present "the old truths in forms that were novel, one might even say in the form of novels."[54]

An effective part of Chesterton's rhetorical strategy is his punning, always light but usually serious too. As in the greatest work of Shakespeare, Milton, and the English metaphysical poets, it is punning with philosophical intent and effect, a technique designed to enliven and involve the reader's common sense, to delight, edify, and remind him of truth, all at once. In St. Augustine's terms, Chesterton would *"docere, delectare, flectere"* — teach, delight, and move his reader. When he wrote that "it is not natural to see man as a natural product," he played dexterously upon the crucially equivocal senses of the word *nature — nature* as metaphysically "fitting," qualitatively distinct from and beyond matter, space, and time, but disposed within them in a way uniquely available to the human mind and spirit; and *nature* as the sum of physical entities, beings consisting primarily of matter within time and the laws that govern and determine them. In calling upon the contrast and conflict between these two senses of *nature*, Chesterton not only cuts powerfully to the crux of the scientism debate, but provides as well a deft and trenchant evocation of an issue that lies at the heart of much of the philosophy, art, and literature of Western civilization, an issue paradigmatically explored in the meditation of the two ideas of nature in Shakespeare's *King Lear*. When Chesterton goes on to say that "it is not common sense to call man a common object of the country or seashore," he does something of the same thing with the word *common* that he had done with *nature*: "the things that are com-

39

mon," such as death or first love, he writes elsewhere, "are not commonplace."

Chesterton always displayed a magnificent skill and thoughtfulness in the manipulation of his language, providing in his well-considered discourse an implicit argument against scientism, which logically contradicts the validity of any coherent thought. But, as we have seen, he also pressed the argument explicitly in no uncertain terms. We do not see man "straight," he indicated (*straight* in the Greek being *orthos*, from which *Orthodoxy*, the title of Chesterton's most famous book, is derived), if we see him as merely an animal: "It is not sane. It sins against the light; against the broad daylight of proportion which is the principle of all reality." And condensed in his use of the term "proportion" is the whole of the orthodox doctrine of *sapientia* and *recta ratio*, or reason rightly used to discern properly the objects and truths of an intelligible universe.

With his Christian common-sense metaphysics, Chesterton fought throughout his life to preserve the traditional concept of man as *homo sapiens* against the threat that science might, as Passmore in his strategically jargon-ridden idiom puts it, "de-anthropomorphize human beings." Sensitive to the increasingly scientistic climate of opinion, he wrote in 1922 that "the creed that really is levying tithes and capturing schools . . . is the great but disputed system of thought which began with Evolution and has ended in Eugenics. Materialism is really our established Church."[55] And in our time, despite having been temporarily discredited by the activities of the Nazis and Communists, eugenics is of course with us again, a fact that Vance Packard has recently and amply documented in his book *The People Shapers*: "In *Homo sapiens*," he quotes biologist R. L. Sinsheimer as saying, "something new appeared on this small globe. The next step for evolution is ours. We must devise that once again on this sweet planet a fairer species will arise."[56] Chesterton entitled his 1922 volume, starkly, *Eugenics and Other Evils*.

The elementary but disastrous confusions of materialism were a favorite object of his satire, harsh or gentle, and he logically thought that the Church would continue to oppose them with its common-sense dualism: "Christianity, which in its early years fought the Manicheans because they did not believe in anything but spirit, has now to fight the Manicheans because they do not believe in anything but matter."[57] His own faith was a rational Christian Theism—"a religion in the sense of a rule; a real trust in some external standard as a reality"[58]—that he was convinced had been lost by many Victorians, their minds in the grip of the idea of progress and the scientism that gave rise to it. He insisted that *homo sapiens*—"the glory, jest, and riddle of the world," in Pope's words—was too great a creature to be "cramped and crushed in that tiny cell that is called the scientific universe."[59] Of scientistic fantasies and religion surrogates such as the nineteenth-century faith in inevitable collective progress—a faith whose credibility was to be weakened somewhat by two world wars, Hiroshima, the gas ovens, and the Gulag—he wrote "I do not like inevitable triumphs";[60] he preferred instead the deliberate and voluntary reforms introduced by ordinary and intelligent citizens. He believed in a present, tangible liberty rather than a prospective theoretical utopia:

> I shall be completely misunderstood if I am supposed to be calling for a return ticket to Athens or to Eden, because I do not want to go on by the cheap train to utopia. I want to go where I like. I want to stop where I like. I want to know the width as well as the length of the world; and to wander off the railway track in the ancient plains of liberty.[61]

And he referred to the great passage in Dante's *Paradiso* (V:19-24) for authority in reasserting the belief that in the proper understanding and use of rational liberty, man was most like the God who endowed him with it.

The tendency of scientistic thinking to destroy the vision of man as *imago Dei* by picturing and perceiving him instead as a creature driven and primarily determined by laws of matter and force was for Chesterton a most vile foolishness. "Something in the evil spirit of our times," he wrote, "forces people always to pretend to have found some material and mechanical explanation" for their own actions and those of other people, when they really know that these are usually the result of "the non-mechanical part of man, the sacred quality in creation and choice."[62] One of the chief causes and culprits was evolutionist thinking, which generated confusions and heresies without end, including a harmfully false revision of the traditional ethical dualism. Popular evolutionism, Chesterton wrote, has

> substituted the Beast for the Devil. It has made us think that our enemy is what they call our 'lower nature,' which means our mere lusts and appetites, things entirely innocent in themselves. The most typical moderns have joined in this. Tennyson, for instance, spoke of moral improvement as 'moving upward, working out the brute'. But was he right? Why should we work out the brute?[63]

In and of itself our animal nature is not bad at all; the source of evil lies in the perverted will, not in the body. Chesterton never shared the stoical, agnostic, Manichean error of regarding the natural being as itself evil—an idea in clear contradiction to the Biblical doctrine of the goodness of God's creation despite its having characterized much of nineteenth-century Puritanism as well as much of the new evolutionist outlook.

On the issue of the physical/metaphysical duality, Chesterton sought balance; matter, he argued, is neither unreal, nor is it the sole reality. But on the other hand, he did insist that within the terms of the duality, the metaphysical retains a precedence over the physical. The idea of "mindless order or objective matter" on their own becoming or producing "subjective mind" is one that takes a far greater credulity than the traditional discernment of design and the inference of a creator, and this fact ought not be allowed to be obscured, Chesterton said, "by the incantation of long and learned words."[64]

The whole host of materialist fallacies derives from the confusion of science with philosophy—or rather the imperialistic arrogation by scientism of the duties, rights, and truths of philosophy. "To mix science up with philosophy," Chesterton writes,

> is only to produce a philosophy that has lost all its ideal value and a science that has lost all its practical value. I want my private physician to tell me whether this or that food will kill me. It is for my private philosopher to tell me whether I ought to be killed.[65]

The materialist cannot without contradiction apply categories of meaning, purpose, value, or ethics, and yet to the extent that his thought and conduct are coherent, they of course depend on these as directive realities. To dismiss these categories out of hand as "merely subjective" is to surrender sanity itself, but the tendency to do so is a chief factor in modern deterministic thinking, leading in practice to demoralization and amoralism. Chesterton sharply criticized such intellectual irresponsibility as "sham science":

> The stupidest or wickedest action is supposed to become reasonable or respectable, not by having found a reason in scientific fact, but merely by having found any sort of excuse in scientific language.[66]

The tendency to "demoralize reason" had begun in the eighteenth century and proceeds apace in our own. It derives from the fallacy of believing that the tradition of reason can be torn from its metaphysical roots and used instead in a merely functional, utilitarian, descriptive way. The sociologists have called this newly truncated use of reason, which grew up especially in the nineteenth century along with industrialism, "functional rationality," and they, among others, have often shown its malignant consequences. The effects of amoral, functional rationality can be seen everywhere in the contemporary world— in the individual, in society, and in the environment; it is the essence of nihilism.

Chesterton's initial realization of the horrors implicit in functional rationality, recorded in his comment about "sham science," provided an insight that gave him, he said, "a momentary illusion of having really got hold of what is the matter with modernity"— the program and attitude of scientism, a "serpent . . . as slippery as an eel," a "demon . . . as elusive as an elf," an "evil and elusive creature."[67] The consequences of a use of reason that claims to be value free and to discern only scientific inevitabilities are underscored in Rabelais's statement about "science without conscience," but they were also pointed out at the end of the nineteenth century by Lord Acton when he noted

> There is a popular saying of Madame de Staël . . . that we forgive whatever we really understand. The paradox has been judiciously pruned by her descendant, the Duc de Broglie, in the words: "Beware of too much explaining, lest we end by too much excusing."[68]

Deterministic thinking eats away at the common reason: scientific thinking starts by scouring away superstitions and falsehoods, but then scientistic thinking, its appetite having grown after continual eating unrestrained by philosophical common sense, ends by devouring truths as well, leaving only the bones and orts of physical reality, with subject and object staring at each other across a yawning faithless chasm.

The obsession with means, with utility, with technique, to the exclusion of a consideration of the ends and purposes that alone can properly regulate and direct them, is the very essence of scientific relativism. One distinguished philosopher has recently attacked it as "the illusion of technique";[69] another has called it nihilistic inasmuch as it proceeds by cause and effect undirected to ends: "the technician," writes Jacques Ellul, "does not tolerate any insertion of morality into his work."[70] Chesterton said that, "like so many modern notions, it is an idolatry of the intermediate, to the oblivion of the ultimate";[71] for example, he noted, we are the age that invented telephones and loudspeakers, and then found that we really had nothing to say—a problem which we "solve" by inventing noisier loudspeakers and better and more numerous telephones. As early as 1905 he became aware of the futility and incoherence of much of our purposeless assault on and "triumph" over nature: "motor-car civilization," he said, "[goes] its triumphant way, outstripping time, consuming space, seeing all and seeing nothing, roaring on at last to the capture of the solar system, only to find the sun cockney and the stars suburban."[72] If we can fault Chesterton for anything here, in words written near the turn of the century, it is for too great a timidity in imagining the enormity of the horrors that would soon occur as consequences of the obsession with means and the denial of ends.

Pursued only as part of that barbaric worship of collective human power that Toynbee saw returning in the eighteenth century from its home in the pre-Christian pagan world, amoral technology has brought us to the point of annihilation. "When the means are autonomous," Charles Williams wrote, "they are deadly."[73] In his *Ends and Means*, Huxley pointed out in 1937 that "It is impossible to live without a metaphysic. The choice that is given is not between some kind of metaphysic and no metaphysic: it is always between good metaphysic and a bad metaphysic."[74] Or, as Chesterton put it, "Men have always one of two things: either a complete and conscious philosophy or the unconscious acceptance of the broken bits of some incomplete and shattered and often discredited philosophy."[75] For Lewis it remained to argue the point conclusively in *The Abolition of Man*.

Empowered by countless enthusiastic disciples and a decades-long reign as the de facto religion throughout a mechanized, materialistic Western civilization, modern scientism has grown in hideous strength to the point that it is capable of quite literally imponderable horrors. The tools are outstripping the toolmaker's capacity to control them; fascination with the means has subverted all consideration of the ends. As Gabriel Marcel wrote a quarter century ago in his aptly titled volume *The Decline of Wisdom*,

> the huge multiplication of means put at man's disposal . . . takes place at the cost of the ends they are supposed to serve, or, if you like, at the cost of the values which man is called upon both to serve and to safeguard. It is as if man, overburdened by the weight of technics, knows less and less where he stands in regard to what matters to him and what doesn't, to what is precious and what is worthless.[76]

As scientism increases, *sapientia* diminishes. No one saw this more clearly, or opposed it more steadily, than G. K. Chesterton. In many contexts and in many ways he reasserted against scientific materialism the essential sanity of Christian common-sense dualism: "[T]he man of science has always been much *more* of a magician than the priest," he wrote; "since he would 'control the elements' rather than submit to the Spirit who is more elementary than the elements."[77] Man is not "a common object of the seashore or the countryside"; he embodies and reveals something unique that draws us beyond all physical, natural categories whatsoever, draws us into the realm of value and meaning, a realm qualitatively distinct from and logically prior to scientific procedures and terms, a realm from which they derive whatever rational coherence, validity, and application they have.

III

Scientism: The Current Debate

I am haunted by the idea that this break in human civilization, caused by the discovery of the scientific method, may be irreparable. . . . The political and military horrors and the complete breakdown of ethics which I have witnessed during my life may be...a necessary consequence of the use of science. . . . If this is so, there will be an end to man as a free, responsible being.

Max Born,
Winner of the 1954 Nobel Prize for Physics

Whether this fear that science might deanthropomorphize human beings has any grounds is the leading philosophical issue . . . of our time.

John Passmore,
Science and Its Critics

*M*AN the knower pursues two related but distinct kinds of knowledge. As *homo sciens*, man the knower of *scientia*, he tends to matters of fact, quantity, matter, and the physical realm; as *homo sapiens*, man the knower of *sapientia*, he shows his interest in the qualities of meaning, purpose, value, idea, and the metaphysical realm. If we are to have truth, neither kind of knowledge can be denied or ignored. The denial of the reality and importance of *scientia* is characteristic of the radical transcendentalism of Eastern religions, but today the even greater and more damaging imbalance is found in the pervasive radical immanentism of much Western culture and thought that attributes validity only to *scientia*. Enthusiasts of scientism fail to see that *scientia* is utterly dependent on *sapientia* for direction and meaning; their fervent attempts to pursue *scientia* in isolation from *sapientia* amount to a tragic rush into meaninglessness—the very antithesis of a genuine search for knowledge.

It was the genius of Chesterton to bring the debate articulately and intelligibly to the attention of the common man, in whom he reposed more trust than in specialists and experts. But as the common man has increasingly fallen under the spell of technology's apparently limitless marvels, the dangers of undirected science and scientistic thinking have only grown in intensity and impact, and the spectre of the destruction of *homo sapiens*—of the abolition of man—can no longer be considered an alarmist nightmare. A generation raised in the spiritual and intellectual vacuum of scientism and taught from birth an abiding contempt for the metaphysical realities underpinning existence has been conditioned to believe that it may well be, as B. F. Skinner argues, beyond freedom and dignity; increasingly, scientistic man finds it no indignity to see and treat himself as what Lewis called "a trousered ape." "The Common Man," Chesterton

48

wrote acidulously, "may well be the victim of a new series of tyrannies, founded on this scientific fad of regarding him as a monkey"[1]—and he made that comment before Nazis, Communists, and other "value free" experimenters gained the power to make his nightmare an historical reality.

In our day, criticism of scientism seems to come most frequently from the scientists themselves. Harvard astronomer and astrophysicist Owen J. Gingerich writes that "scientism is a . . . dogmatic philosophy that can develop from [scientific observation], saying that since this is the only way we can find out about nature, that is all there is."[2] On the other hand, part of the reason that some scientists are reluctant to accept the Big Bang theory of the origin of the universe, writes the physicist Robert Jastrow, Director of the United States National Aeronautics and Space Administration's Institute for Space Studies, is that the theory is inconsistent with the religion of science: "This religious faith of the scientist is violated by the discovery that the world had a beginning under conditions in which the known laws of physics are not valid, and as a product of forces or circumstances we cannot discover."[3] Yet with the "de-materialization of matter" brought about by the insights of modern physics (especially those of Einstein and Heisenberg), it is safe to say that large numbers of intelligent scientists have moderated the bad metaphysic of reductive materialism and the accompanying hubris that they inherited from many of their nineteenth-century predecessors.

In the contemporary world, as the philosopher Richard Kroner wrote some years ago, it is often true that "scientism is the vice of those who do not practice science itself but are intoxicated by the triumph of scientific, and even more of technological, discoveries and devices, i.e., the vice of the masses in almost all countries on the earth."[4] More recently the philosopher Henryk Skolimowski has noted that

> Present science does not lend much support to what is popularly called the scientific-technological world view. Yet this world view, although it has been critically undermined by science itself, is powerfully sustained by a variety of institutions, including present schools and academia. What is it, then, that gives this world view its legitimacy and sustaining force? Above all, it is our ideals of secular

49

salvation, which have generated all the welter of socio-economic institutions, including the motor industry and habits of conspicuous consumption.[5]

Nevertheless, despite a shaky philosophical foundation and criticism from both within and without the scientific community, reductionism continues to command widespread allegiance. Certainly one reason for the Eastward pilgrimage that so many young Westerners, and especially Americans, are taking today is their boredom and disgust with minds and institutions that exhibit what Kierkegaard called "the inability to rise above the absolute reality of time and space," the dogmatic world view of scientism. In America the problem has been intensified by a development that the educational philosopher Philip H. Phenix noted as early as 1955, when he wrote that in the American public schools "it seems unfortunately to be the case that what has been presented as a means for preserving religious peace and freedom through secularization has to some extent become a method of propagating a particular dogmatic faith, namely, scientific naturalism or to give it another name, naturalistic humanism."[6]

Since 1955 this trend has become ever more solidly entrenched within America, and the Western world has been, in this regard as in others, for good and ill, Americanized. The resulting educational representations of the world have been, of course, largely unconvincing; as E. F. Schumacher put it in the last book he wrote,

> The maps produced by modern materialistic Scientism leave all the questions that really matter unanswered; more than that, they deny the validity of the questions. The situation was desperate enough in my youth half a century ago; [but] it is even worse now because the ever more vigorous application of the scientific method to all subjects and disciplines has destroyed even the last remnants of ancient wisdom—at least in the Western world.[7]

The increase of scientism necessitates the decline of *sapientia*, of coherent rational and philosophical views. The "scientized reality principle," writes Theodore Roszak, ". . . treats quantities as objective knowledge and qualities as a matter of subjective preference."[8] Out of despair or contempt with the result comes much of the cultural fragmentation around us, a fragmentation that is itself the effect of revulsion against the tyranny of quantity and the imperial triumph of *homo sciens*.

The breakdown in the transmission of general philosophical culture has helped make the industrialized, scientized "Wasteland" of T. S. Eliot's imagining a worldwide phenomenon, at immense personal and spiritual—and environmental—cost. This is of course even more true in states where no criticism of the scientistic world view is allowed, the domains of "scientific socialism." (John Strachey, writing in 1936, asserted that the Communist Party was "the only body of persons who possess a knowledge of the science of social change," thus describing precisely the illusion of scientism applied to politics that has cost millions of lives in our century.) Yet even in the Soviet Union, the concrete pavement of scientific materialism has cracked in some places, and shoots of criticism of "the illusion of technique" have come forth "from under the rubble."[9] Solzhenitsyn's play "Candle in the Wind" is an explicit and detailed philosophical attack on scientific materialism and the manipulative "technobureaucratic elite" who are its prime beneficiaries. This "new class," first pointed out as such by the Yugoslav Milovan Djilas,[10] is a new priesthood that accumulates more and more power as the staggering falsehoods and errors of "scientific socialism" go uncontradicted in official Soviet and other Communist cultures.

It is important to remember, however, that Western utilitarianism has the same eighteenth- and nineteenth-century roots and is characterized by many of the same ideas and developments as those in Communist cultures. Sociologist David Martin has illuminated this similarity of development by showing how what he calls "scientistic ideology" erodes the legacy of metaphysical common sense that recognizes in life, society, institutions, and culture a "qualified dualism," which he says is "symbolized either in debate and seminar, or else in a series of distinctions between church and state, sacrament and material world, body and soul, priest and administrator." Under the powerful assault of scientistic ideology, Martin suggests, this dualism is broken down, "simplified," reduced: "Debate must give way to technical committee, seminar to laboratory, and the office of administrator can be merged with that of priest, who then becomes a scientific coordinator." The general trend of this development is that "Religion and politics are both assimilated to science," and "[j]ust as there is no disagreement in science there can be no disagreement in society: hence the government of people may give way to the administration of things."[11]

And it is in an important sense the ultimate effect of scientism to dissolve the absolute qualitative distinction between persons and things—the very heart of the metaphysical tradition, of *sapientia*—reducing persons *to* things, denying man's rational soul and his transcendence of the physical, giving him a value no higher than that of a camel or a stone or any other part of nature. This reduction of the human category to the natural runs parallel with a whole series of reductions from quality to quantity, from value to fact, from rational to empirical. If the doctrine of man as rational moral being, qualitatively distinct from and incommensurate with nature, is weakened or destroyed, the grounds for expecting or encouraging moral conduct are similarly weakened. What reason, Whitehead has asked, could men like Hume or Thomas Huxley have given for *any* moral views they held, "apart from their own psychological inheritance from the Platonic religious tradition"?[12] Yet in common-sense morality, even where simple politeness exists, there is always some implicit recognition of people as ends and never only means, some recognition that, as Rebecca West once put it, "perhaps the sin against the Holy Ghost is to deal with people as though they were things." This remnant of traditional piety undoubtedly provides for some people the only scrap of traditional *humanitas* that supports conventional morality.

It is important to recall, however, that the case against scientism extends beyond its malignant and demoralizing effects, important as these are, to the internal contradictions that make it not only, in Aldous Huxley's terms, "a bad metaphysic," but also a false and irrational one. Ever since the seventeenth century, philosophers and metaphysicians have pointed out these contradictions, but it does not take a philosopher to understand them; common sense at its best can recognize and point them out, spurred on only by a dogged commitment to rational consistency, to the realization that the scientific method derives from the rational method, and not vice versa. Too often we are blinded to such common-sense conclusions by a tragic loss of faith in our own rationality, by our willingness to surrender the distinctions between means and ends, persons and things, value and fact, mind and matter, subjects and objects, truth and falsehood, consistency and contradiction, whenever scientific language is used. We are intimidated and silenced by jargon. "Most of the machinery of modern language is labor-serving machinery," Chesterton wrote;

> and it saves mental labour very much more than it ought. Scientific phrases are used like scientific wheels and piston-rods to make swifter and smoother yet the path of the comfortable. Long words go rattling by us like long railway trains. We know they are carrying thousands who are too tired or too indolent to walk and think for themselves.[13]

Long words should not be allowed to hide elementary rational contradictions. Judgments of value are made at the outset of any scientific study or experiment, whether this is apparent to the student or experimenter or not. "There has been conscious selection of the parts of the scientific field to be cultivated," Whitehead says,

> and this conscious selection involves judgments of value. These values may be aesthetic, or moral, or utilitarian, namely, judgments of exploring the truth, or as to utility in the satisfaction of physical wants. But whatever the motive, without judgments of value there would have been no science.[14]

Science is derived from the rational method of philosophy and is dependent on it for estimates as to the meaning and value of what is proposed, observed, or discovered. Objectivity itself is a judgment of value: the old and hallowed value of careful impartiality or disinterestedness in observation and judgment. It is not an immediate or necessary inference from any "objects" per se, but the resulting confusion of "objects" and "objectivity" does great damage by obscuring the attributions of value that are made in every choice, decision, experiment, or selection.

"When I make a statement, even as coldly and impersonal a statement as a proposition of Euclid," it has been written, "it is I that am making the statement, and the fact that it is I that am making the statement is part of the picture of the activity." And

P. W. Bridgman continues by saying that "[t]he insight that we can never get away from ourselves is an insight which the human race through its long history has been deliberately, one is tempted to say willfully, refusing to admit."[15] This is not necessarily a counsel of despair or skepticism, or a warrant for subjectivism: far from it. It is the invitation and the demand to raise the issue to the philosophical plane and discuss it in terms of validity: it is the metaphysical issue made explicit. The bad faith and sheer incoherence of refusing this invitation, of not responding to this command, are at the heart of the scientistic ideology, at the root of scientific materialism, and they lead inevitably to contradiction and more and more often, in today's world, to disaster.

At least sometimes Charles Darwin himself seems to have recognized the disjunction of these realms, the qualitative and the quantitative, the rational and the empirical, the philosophical and the scientific: as a perpetual reminder he kept a piece of paper on which he had written "never use the word higher or lower"—i.e., qualitative language—when talking about evolution by natural selection. This is the kind of confused naturalistic reductionism of which the obvious next step—and fallacy—was articulated by Enrico Fermi when he said that "whatever Nature has in store for mankind, unpleasant as it may be, men must accept." Fermi was one of Max Born's students, and it was precisely Fermi's kind of incoherent reductionism that Born was to deplore in later years: "They were such clever and efficient pupils," he said of Fermi and some of the others who later made the atomic bomb, but Born felt guilty that somehow "all they learned from me were methods of research"; he wished, he later wrote, that "they had shown less cleverness and more wisdom"—that is to say, less scientism and more *sapientia*.[16]

In recent years the philosopher Hans Jonas has succinctly and intelligibly illuminated the typical root contradiction of the scientistic ideology: "Modern theory is about objects lower than man," he writes; "even stars, being common things, are lower than man." Yet even in the so-called "human sciences," whose object of attention and study *is* man, Jonas notes, the object remains lower than man: "For a scientific theory of him to be possible, man, including his habits of valuation, has to be taken as determined by causal laws, as an instance and part of nature." Here is Bronowski's assertion again; man is reduced and completely assimilated to the realm of quantity, matter, and force. But Jonas points out the inevitable and decisive contradiction: "The scientist does take [man to be determined by causal laws]—but not himself while he assumes and exercises his freedom of inquiry and his openness to reason, evidence, and truth." His own working assumptions involve free will, deliberation, and evaluation as aspects of himself, but those qualities and capacities are stripped away from and denied to the human "object" or "thing" that he is inspecting. Often unwittingly engaged in, this is the classic reductive process, what Koestler calls "the ratomorphic fallacy" which, in Chesterton's terms, "treats man as a monkey." Hans Jonas describes the result:

> man-the-knower apprehends man-*qua*-lower-than-himself and in so doing achieves knowledge of man-*qua*-lower-than-man, since all scientific theory is of things lower than man-the-knower. It is on that condition that they can be subject to "theory," hence to control, hence to use. Then man-lower-than-man explained by the human sciences—man reified—can by the instructions of these sciences be controlled (even "engineered") and thus used.[17]

This is indeed treating man as a "thing," as a "common object of the countryside," as a part of nature just like a camel. It is not only an inhumane procedure, it is simply false according to ordinary standards of reason, although it is no less widespread for that. It is the great modern religion, our established church, with a whole panoply of priests, evangelists, saints, and bishops, and massive means of publicity and propaganda. It has the power, if it is allowed to grow uncontested in enough human minds, to bring about the end of *homo sapiens*. It may indeed prove to be the abolition of man.

IV

C. S. Lewis and the Two Cultures

When science has discovered something more
We shall be happier than we were before.
Hilaire Belloc

The scientists have the future in their bones.
C. P. Snow

*T*HE eloquent style and superbly balanced reason that Lewis used in making the case against scientism and, conversely, in making the case *for* man as *homo sapiens* have earned him an extraordinary popularity. The very fact that he chose to take a stand on these controversial issues in their explosive heyday has assured his importance in the modern history of ideas. And yet, neither his assertion concerning the nature and character of man nor his critique of scientific materialism is by any means unique or original, as we have seen; on the contrary, his interest lay not in the manufacture of a brave new world, but with the preservation of the old verities, the revitalization of an age-old philosophical tradition, the restitution of man. What is unique in Lewis is his genius for having addressed his arguments to the common man as he did, pointing out the immeasurable worth of man in a manner neither trivial nor obsequious, but with a cool, clear, and reasoned appeal to common sense. Whatever the profundity of his thought—and it is considerable—he always took care to make it wonderfully penetrable.

Lewis's use of popular idiom to clarify issues proved a particularly effective strategy in dealing with scientific materialism, because the most zealous of the apostles of scientism (and of its concomitant naturalistic reductionism) are to be found not so much among the professional elite as among the nonprofessional enthusiasts of science—the propagandized common man. Chesterton wrote brilliantly in this mode as well, although to less eventual effect, since his wit and easy journalistic manner led many to underestimate the genuine metaphysical substance of his thought. A more extensive background in philosophy gave Lewis the intellectual depth necessary to fill out Chesterton's magnificent intuition while retaining something of his stylistic elegance and dexterity.

In Lewis one can hear the great clerkly tradition of *recta ratio* that in England has been transmitted through Cambridge (Milton and the Cambridge Platonists) and through Oxford (Richard Hooker, Samuel Johnson, and John Henry Newman). It is the great central tradition of the "English Moralists" of whom Basil Willey has written—great and central not merely because it is English, but because it draws on the resources of the continental tradition of rational Christian Theism, of Erasmus and Aquinas and the Scholastics, who mediated Augustine, the Fathers, the Scriptures, and Plato and Aristotle to Medieval and Renaissance England. It is a tradition that has maintained a balance between two extremes—an excessively transcendental, stoical Christianity that denies the goodness of nature (e.g., Calvinism and the Puritan tradition), and an excessively empirical tradition that tends to deny all but a strictly naturalistic, material reality.

In recent generations extreme naturalism has been the most pervasive and malignant excess; as an Oxford philosopher put it over fifty years ago,

> The philosophy current to-day is almost without exception riddled with fallacies arising out of the uncritical application to philosophical questions of methods and results derived from the sciences; and the progress of modern philosophy is intimately bound up with its liberation from a scientific dogmatism which has long ago exhausted its contribution to philosophical thought.[1]

When Collingwood leveled these charges in 1924, Lewis, having already earned a First in English, and a Double First in Classics, was tutoring in Philosophy at University College, Oxford. Despite his subsequent decision to make literature rather than philosophy his vocation, Lewis's philosophical education continued to give extraordinary substance, depth, and logical power to all his writing, and he used that education throughout his life in attacking precisely those fallacies and that scientific dogmatism that Collingwood abhorred.

But in Oxford and in the rest of the Western academic

community, scientific naturalism was to have nearly as great an effect on philosophy departments as it had on the world at large. The extreme naturalism and nihilism of A. J. Ayer as presented in his *Language, Truth and Logic* constituted what was perhaps the nadir of serious Anglo-American philosophy in this century. Father Martin D'Arcy told Evelyn Waugh that Ayer was "the most dangerous man in Oxford," and so he was: his thralldom to scientism was complete. As Ayer himself put it recently, "the scientists felt *Language, Truth and Logic* was all right. After all, it told them they were the most important people, and they liked that." When asked what were the "real defects" of his Logical Positivism, Ayer replied: "I suppose the most important of the defects was that nearly all of it was false."[2]

The case against traditional ethics that Ayer made and the attitude toward reality that he proposed were Lewis's chief targets in *The Abolition of Man*, his own manifesto for the common man as well as the philosopher against aesthetic nihilism. In order to have an accurate sense of the challenge Lewis seized, and of his accomplishment, it is important to have some understanding of the kind of professional fecklessness and learned idiocy he opposed—the sort of thing Tom Stoppard satirized in his play *Jumpers*, in which a character aptly notes that "modern philosophy has made itself ridiculous by analysing such statements as, 'This is a good bacon sandwich,' or, 'Bedser had a good wicket.'"

The knowledge of things and of the means by which to manipulate them most effectively was the goal of Bacon's "great instauration" in the seventeenth century, and it is the central defining feature of modern technology, although more and more intelligent modern people have grown wary of the brutal simplicity of the goal of "effecting all things possible." The moral and rational vacuum at the heart of utilitarianism—Marxist and revolutionary or liberal and evolutionary—is in its utterly inadequate conception of validity and obligation, its attenuated grasp of any coherent concept of a good other than pleasure; yet it is the only ideology behind the modern religion of technique. The American Jesuit John Courtney Murray is only one of many who has pointed out that "the only canon of technology is possibility," and no form of utilitarianism can coherently or effectively check this nihilistic tendency, since the only criterion coherently proposed by utilitarianism is the pleasure/pain calculus, of which each person is obviously the only accurate and necessary judge in his own life. Such premises imply nothing short of complete ethical relativism.

Perhaps its ethical implications are the reason Heidegger wrote that "the essence of technology is danger," and his disciple the distinguished American philosopher William Barrett goes on to explain that "technology, when it becomes total, lifts mankind to a level where it confronts problems with which technical thinking is not prepared to cope." Discussing the issue of eugenics, to which Chesterton and Lewis both gave attention, Barrett writes that "technique by itself cannot determine a philosophy . . . the powers of genetic manipulation, were they all at our disposal, would not provide the wisdom for using them. That must come from another kind of thinking, for which a technical civilization might have become incompetent through sheer lack of practice."[3]

The development of expertise (*scientia*) in the accumulation and manipulation of technical power, without a corresponding development of knowledge (*sapientia*) as to the right uses, purposes, goals or values which that power ought to serve, is the distinctively modern form of science sans conscience and leads to the same dehumanization and barbarity. "Modern man's state of mind is completely dominated by technical values," Jacques Ellul wrote in his classic *The Technological Society* in 1954, "and his goals are represented only by such progress and happiness as is to be achieved through techniques."[4] These are the criteria of power and pleasure without the criterion of goodness, with the result being the use of a merely functional rationality to accumulate what Lewis called "that hideous strength."

"I agree Technology is *per se* neutral," Lewis wrote to Arthur C. Clarke in 1943: "but a race devoted to the increase of its own power by technology with complete indifference to ethics *does* seem to me a cancer in the Universe."[5] To this succinct summary of the problem of scientism we can add F. R. Leavis's assessment made two years later in *Education and the University*: "On the one hand there is the enormous technical complexity of civilization, a complexity that could be dealt with only by an answering efficiency of coordination—a cooperative concentration of knowledge, understanding, and will (and 'understanding' means not merely a grasp of intricacies, but a perceptive wisdom about ends)." Here is a plea for *sapientia*; however, Leavis too sees that the "decline of wisdom" has made this harder and harder: "on the other hand, the social and cultural disintegration that has accompanied the development of the inhumanly complex machinery is destroying what should have controlled the working." This had been Blake's intuitive and prophetic insight too—and provided the animus for his attack on empiricism—"mind-forg'd manacles" and "dark satanic mills," which is to say mechanism in thought and economic development. "It is as if society," Leavis continues, "in so complicating and extending the machinery of organization, had incurred a progressive debility of consciousness and the powers of co-ordination and control—had lost intelligence, memory, and moral purpose."

The American Humanists Paul Elmer More and Irving Babbitt, and one of their students and friends, T. S. Eliot, were also making this case in the 1930's against what Leavis called "Techno-Benthamism" and what Lewis called "futilitarianism." Eliot was to trace its psychological and philosophical roots to the seventeenth century and explain them by means of the concepts of secularization and the "dissociation of sensibility," the splitting of mental integrity into subjectivist and objectivist extremes, Rousseauvian and Baconian; and Herbert Butterfield in *The Whig Interpretation of History* showed how the ideology, the scientistic faith in collective human progress, gradually displaced the belief in the primacy of moral responsibility and personal progress that lies at the heart of the traditional religious and philosophical view of the world.

This is the general background to the "two cultures" dispute that emerged explicitly around 1960 in the literary debate between F. R. Leavis and C. P. Snow, with the latter accusing teachers and students of the humanities of being "Luddites" in their mistrust and criticism of science and technology. It should be evident that Lewis shared this mistrust with Leavis, but it is important to note the kind and extent of their agreements and disagreements on this and related issues. One practical way of doing this is simply to describe the way in which each of them dealt with the legacy of the humanistic critique of science that had been made in the nineteenth century by Matthew Arnold, to whom Leavis was largely, and Lewis somewhat less, indebted for attitudes and ideas about literature, religion, science, and life.

In *Culture and Anarchy*, Arnold had described and criticized the moral and spiritual condition of England in the mid-nineteenth century: "the whole civilization is, to a much greater degree than the civilization of Greece and Rome, mechanical and external, and tends constantly to become more so." The commercial, the quantitative, and the mechanical increased apace, brought about by "our worship of machinery, and of external doing."[6] Here is the "illusion of technique" that accompanied and helped fuel both industrialism and the new religions of humanity, and their chief dogma, the belief in a poorly defined and shallow but "inevitable" collective progress. But, like Goldsmith before him, Arnold saw that "wealth accumulates but men decay," that it ill profits a man to gain power over the whole natural world if thereby he loses the belief in, and the possession of, the reality of his own soul: "The idea of perfection as an *inward* condition of mind and spirit is at variance with the mechanical and material civilization . . . nowhere . . . so much in esteem as with us."[7]

Both Lewis and Leavis—as well as the Americans Babbitt, More, and Eliot—agreed with this critique and continued to refine and apply it in their own writing, and all of them also saw in humane literature a precious repository of wisdom that could help people to realize their true condition and free themselves of the illusions of scientism. Language itself, Leavis wrote, "is a vehicle of collective wisdom and basic assumptions, a currency of criteria and valuations, " and "it entails on the user a large measure of accepting participation in the culture of which it is the active living presence."[8] This penetrating grasp of the assumptions and implications of language as the uniquely human and moral aspect of our mental life and social life is of immense value and importance and shows Leavis at his best, having "a mind in possession of its experience" with "a considered scale of values."[9] Like Lewis and Chesterton and Newman, Leavis hearkens to the profound ethical and aesthetic common sense of that great concentric sage Dr. Johnson in contradistinction to eccentric and excessive subjectivism and individualism: "Johnson is not," Leavis wrote, "like the Romantic poet, the enemy of society, but consciously its representative and its voice, and it is his strength—something inseparable from his greatness—to be so."[10]

Reading, studying, and teaching great literature was for Leavis the chief form of moral activity and the most vital feature of a person's education ("The questions work and tell," he wrote, "at what I can only call a religious level").[11] This was all the more important because of the growing barbarism and amorality brought about in the modern world by scientism: "My own recourse to the word 'spiritual,'" he wrote in *Lectures in America*, "is determined by the contemplation of a world in which the technologico-Benthamite ethos has triumphed at the expense of the human spirit—that is of human life."[12] For Leavis, novelists such as George Eliot and Henry James had religiously incarnated value, meaning, and purpose of the highest order in their works, as had poets like Ben Johnson and Pope. In discussing Pope's moral critique of the conditions and values of eighteenth-century England, Leavis said "his contemplation is religious in its seriousness."[13] In the light of this faith and vision came Leavis's contempt for what he rightly perceived as the aesthetic nihilism of much of modernism and especially of Bloomsbury: "Articulateness and unreality cultivated together; callowness disguised from itself in articulateness; conceit casing itself safely in a confirmed sense of high sophistication; the uncertainty as to whether one is serious or not taking itself for ironic poise."[14] He said of the Sitwells that they belonged more to the history of publicity than to the history of poetry.

Thus Leavis fought what he considered two barbarous extremes that were destroying society—the scientism of the "technologico-Benthamite ethos," a form of ideological reductionism masked as objectivity exemplified by utilitarians such as C. P. Snow; and the nihilism of the aesthete, a form of feckless impressionism, subjectivism, and hedonism, exemplified by such artists as Virginia Woolf—to both of which he preferred as antidote the great tradition of moral imagination in literature. The author of a recent book on the preoccupations of Leavis's literary criticism argues that "the principle underlying [those preoccupations] is not moral but religious"; and another recent writer asserts that "'the religious' for Leavis is what *preserves meaning* in a world of technologico-Benthamite reductivist materialism."[15]

64

Much as Lewis agreed with Leavis on almost all of the foregoing, it was on this question of the nature and role of religion that he disagreed, and the disagreement is rooted in and illustrated by Leavis's indebtedness to Arnold's conception of religion. Of course Lewis and Leavis disagreed about other things too, especially about the continuing value of Milton, for whom Lewis unabashedly defended his love: "It is not that [Leavis] and I see different things when we look at 'Paradise Lost'. He sees and hates the very same things that I see and love."[16] Underneath what seems to be a merely aesthetic disagreement is the real root of the antagonism: with Dante and Spenser, Milton stands among the great and explicitly theological poets, and it is over theology that Lewis and Leavis parted ways.

Leavis's effort, like Arnold's, was bent toward constructing a "religion of culture," a religion without a theology and without a God. Arnold foresaw poetry gradually replacing religion; the reading, teaching, and study of literature, he was certain would somehow replace religious doctrines and institutions as the center of moral authority, value, and culture. Arnold's much-esteemed teacher Newman had presciently seen and denounced the coming of this liberal humanistic religion—whether based on poetry or science or both—as early as 1841: "If we attempt to effect a moral improvement by means of poetry, we shall but mature into a mawkish, frivolous, and fastidious sentimentalism—if by means of argument, into a dry unamiable longheadedness—if by good society, into a polished outside, with hollowness within, in which vice has lost its grossness, and perhaps increased its malignity—if by experimental science, into an uppish supercilious temper, much inclined to scepticism."[17]

In his history of *The English Critic*, James Sutherland notes that it was precisely a religion of culture that Arnold effectively recommended: "Arnold's Puritanism is not of the sixteenth- or seventeenth-century kind which tends to look upon all imaginative literature with disapproval." Instead, Sutherland writes, Arnold's "Puritanism takes the curiously inverted form of giving to literature a disproportionate importance."[18] With this assessment Lewis agreed. "The present inordinate esteem of culture by the cultured began, I think, with Matthew Arnold—at least if I am right in supposing that he first popularized the use of the English word *spiritual* in the sense of German *geistlich*."[19] Yet no one loved literature—or more effectively communicated that love in writing or speaking—than Lewis. "A Lewis lecture was a feast," Kingsley Amis has written; "if ever a man instructed by delighting it was he"; he was "a masterly teacher and a critic whose knowledge and feeling were in unusual accord."[20] But Lewis was too well versed in philosophy to fail to recognize the dangers and fallacies of even a stoical aestheticism, no matter how sincerely and strenuously ethical its individual votaries might happen to be—and he knew that men such as Babbitt and Leavis and Trilling were strenuously and passionately good and moral men. He agreed with Newman, and with the great medieval scholar Curtius whom he quoted as saying that "the modern world immeasurably overvalues art." All previous ages were right, Lewis thought, "in making art . . . subordinate to life."[21]

Eliot too made this criticism of the tendency of Arnold's thinking as seized and applied by his own teacher and friend Irving Babbitt, a literary moralist much like Leavis. Refuting Arnold and Babbitt, Eliot called upon Jacques Maritain's assertion that "It is a deadly error to expect poetry to provide the super-substantial nourishment of man"; and Eliot himself wrote that "it is ultimately the function of art, in imposing a credible order upon reality, and thereby eliciting some perception of an order *in* reality, to bring us to a condition of serenity, stillness, and reconciliation; and then leave us, as Virgil left Dante, to proceed toward a region where that guide can avail us no farther."[22]

Lacking a religion beyond literature, Leavis was often forced into a premature and excessive moralism within literature, creating a canon of works in "the great tradition" and excluding or excommunicating from it books or writers who failed to measure up to his criteria, which were chiefly "reverent openness before life and a marked moral intensity." But the requirement of a direct and didactic moral value and correlation in the literary work drastically foreshortens and reduces aesthetic autonomy and human variety. In literature, as Lewis put it, "we seek an enlargement of our being," but in order for this to be possible we must allow latitude to art—we ought to give a greater aesthetic autonomy and sympathy to works of art than Leavis and "the vigilant school of critics" will allow.[24] For Eliot and Lewis, and for

66

W. H. Auden, who was profoundly influenced by both, only if art is held to be subordinate to both life and religion can it truly be freed to do its work of enlightening or entertaining or inspiring.

Interestingly, there is one aspect of Arnold's stated philosophy to which both Lewis and Lionel Trilling were able to appeal in criticism of Leavis's stoical religion of culture. "Criticism is for him," Lewis writes of Arnold, " 'essentially' the exercise of curiosity, which he defines as 'the disinterested love of a free play of the mind on all subjects for its own sake'. The important thing is 'to see the object as in itself it really is'."[25] Much literary criticism ought, then, to be simply description and understanding, and there should be a suspension of final judgment until the describing is well and truly done; "the greatest cause of verbicide," Lewis wrote, "is the fact that most people are obviously far more anxious to express their approval and disapproval of things than to describe them."[26]

The point Lewis makes is rooted in his understanding of the philosophical problem of validity, and he has in mind not only what he takes to be the often excessively moralistic criticism of Leavis, but also, and far more deeply, the debunking of the objectivity of any and all value statements conducted by such as A. J. Ayer with his theory of emotivism—the contention that when any one says "X is good" he is saying only "I like X." Assertions of personal preference disguised as assertions of value not only provide grist for the debunker's mill, but by doing so they weaken the conception of validity itself. Leavis's sincere and earnest sense of religious responsibility, like that of Arnold and Babbitt, fell short of philosophical consistency, and both Eliot and Lewis knew this; Leavis criticized the errors of naturalism without following out to its logical conclusion the criterion by which he did so. But it is important to remember that for all their disputes, Lewis thought Leavis's critique of technologico-Benthamism was not only sincere but right and rightly directed, and that he was a man of passionate moral striving and good intention.[27]

With the question of ultimate values we enter the terrain on which Lewis knew the decisive battles for meaning and truth would have to be fought, battles for which the impressionistic and amateur theologies and philosophies of Arnold and Leavis (and of Irving Babbitt and Lionel Trilling, for that matter), however sincerely well-intentioned, were inadequate. It is not enough to talk vaguely, as Arnold did, about religion as "morality touched by emotion" and about God as "a power external to . . . ourselves that makes for righteousness."[28] Arnold's moral certitude concerning the obligation and importance of every man to try to "see life steadily and see it whole," to adopt "the disinterestedness which culture enjoins, and its obedience not to likings and dislikings," and "to enlarge our first crude notions of the one thing needful," is expressed in the vocabulary of Christian morality; but such a system of moral imperatives cannot stand or flower without the roots of rational Theism and of that *recta ratio* which he himself commends but did not consistently follow out.[29] He shied away from Newman's use of "illuminative reason" and moved toward a liberal, aesthetic Christianity not much different in content from pious stoical agnosticism, and in this regard he was very much influenced by the liberal age and the progressive climate of opinion in which he lived, an age that ended conclusively and catastrophically in 1914.

Lewis's defense of reading and literature is no less passionate than that of Arnold or Leavis, but he grounds it in far deeper and more coherent arguments, arguments that reach and illuminate deftly but profoundly the philosophical realm itself. "Good reading" he wrote in a great passage toward the end of *An Experiment in Criticism,*

> though it is not essentially an affectional or moral or intellectual activity, has something in common with all three. In love we escape from our self into one other. In the moral sphere, every act of justice or charity involves putting ourselves in the other person's place and thus transcending our own competitive particularity. In coming to understand anything we are rejecting the facts as they are for us in favour of the facts as they are. The primary impulse of each is to maintain and aggrandise himself. The secondary impulse is to go out of the self, to correct its provincialism and heal its loneliness. In love, in virtue, in the pursuit of knowledge, and in the reception of the arts, we are doing this. Obviously this process can be described either as an enlargement or as a temporary annihilation of the self. But that is an old paradox: 'he that loseth his life shall save it'.[30]

Here is eloquently and concisely distilled a lifetime of philosophical, ethical, and aesthetic reflection, all tied together and made coherent by that rational and religious sense, that "illuminative reason" which Newman had aptly identified as "not only a portion, but a condition of general knowledge." This is "seeing life steadily and seeing it whole," realizing the golden mean between the extremes of what Whitehead has defined as "barbaric vagueness" and "trivial order," occupying and articulating the humane space between them; it is what Quintilian defined as "the good man speaking well"; it is *sapiens atque eloquens pietas*—a wise and eloquent piety—which Babbitt outlined in 1908 as the virtue that the classical-Christian educational project always sought to produce.[31]

In an era of widespread cultural and intellectual specialization and fragmentation, and of emotional and personal alienation and dissociation, Lewis achieved and maintained an "undissociated sensibility," as Kingsley Amis suggested by saying that Lewis's "knowledge and feeling were in unusual accord." Lewis sees "in love, in virtue, in the pursuit of knowledge, and in the reception of the arts" the functioning of the separate but related and distinctively human activities of *homo sapiens*, man as the participant, alternately, in the recognition, the enactment, and the attribution of value itself. Reading provides a means of access to this realm of *sapientia*, self-knowledge, and humane experience, although of course not necessarily the only means:

> in reading great literature I become a thousand men and yet remain myself. Like the night sky in the Greek poem, I see with a myraid eyes, but it is still I who see. Here, as in worship, in love, in moral action, and in knowing, I transcend myself; and am never more myself that when I do.[32]

When the self transcends its competitive particularity it is sapient; it both knows and tastes truths beyond the ego; it is disinterested rather than self-indulgent. Lewis achieved a genuine understanding of value and obligation—inchoate in Arnold and Babbitt and Leavis—and placed it within a meaningful context by recovering and clarifying their philosophical and theological bases and purpose. Not out of mere hyperbole does the Bible say, "O taste and see, that the Lord, He is Good," for meaning and value, when properly and genuinely assimilated, are part of the sacramental nature of reality which encompasses and supercedes both the aesthetic and the ethical realms and modes of knowledge and experience.

A key to understanding Lewis's assertions concerning the utility of literature in the pursuit of *sapientia* can be found in his comparison of reading with worship, in both of which he recognizes a profound depth of meaning. An expert and incisive etymologist, Lewis was attuned to the root meaning of *worship*—literally, the state or condition of worth—with its correlative recognition, assimilation, celebration, and proper attribution of value itself as a metaphysical reality distinct from any material object to which it might be descriptively applied. Worship is for him categorically related to love, to moral action, and to knowledge in its function as a means by which the individual can transcend both the physical realm and the narrow bounds of ego; indeed, worship is the fundamental structure for all the forms of transcendence, all the paths to *sapientia*—it is, borrowing Newman's words once again, "not only a portion, but a condition" of all true realizations of value.

There is more to say concerning Lewis's articulation of this premise in *The Abolition of Man*, but the point can be briefly summarized. There is an unconditional value that man among earthly creatures is uniquely fitted and obliged to recognize, to apprehend, to imitate, to enact, and to use as a standard and criterion from which all of his subordinate valuations derive, whatever accuracy and truth they have: it is the conviction, in Lewis's words, that "good is indeed something objective and reason [is] the organ whereby it is apprehended."[33] This is the conviction of the *philosophia perennis*, the great central tradition of Western philosophy, metaphysics, religion, and ethics. This objective good can be apprehended, although never totally comprehended, in this life because it is rooted in eternity and the sapience of God; it is always more than, but never less than, what human wisdom at its careful best necessarily infers. As Michael Dummett, the new Wykeham Professor of Logic at Oxford, put it, with admirable clarity and brevity, "I find no satisfactory account of truth or ethics without Theism." Lewis and T. S. Eliot, both of whom were trained philosophers, reached the same conclusion.

This is the core of rational Theism—Jewish, Christian, or Moslem—that a man can apprehend the Good and try to live by it. It does not bring philosophy, science, literature, and the other realms of human culture and activity grinding to a halt with a grand climactic resolution of all problems. We see through a glass darkly, but we do see enough to fare forward—the "we" including every person of good will and honest reflection. Faith in the rational Good stands firm between extremes of premature closure and narrow-mindedness on the one hand, and interminable indecision and "broad-mindedness" on the other. It serves today as it once served Addison, whom Dr. Johnson described as having a disposition of mind which was a precondition not only for logical and religious knowledge, but for *any* accurate and true knowledge: his faith had in it nothing either "weakly credulous, or wantonly sceptical."

In an era when, as one contemporary cultural historian puts it, "the idiosyncratic has triumphed over the normative,"[34] the feckless specializations and fragmentations of modern culture that are the by-products of scientism have driven this kind of *sapientia* into retreat and wittingly or unwittingly have tried to exterminate exalted common sense, because of its having provided an uncompromising and unflattering witness and reproach to their own fecklessness and futility. Defending what he called "that indispensable equipment we call common sense," Jacques Maritain wrote in one of his last books that "when everyone starts scorning these things, obscurely perceived by the instinct of the spirit, such as good and evil, moral obligation, justice, law, or even extra-mental reality, truth, the distinction between substance and accident, the principle of identity—it means that everyone is beginning to lose his head."[35]

Maritain was writing angrily at the end of the 1960s, a decade filled with fashion, fad, and fraud in the intellectual and cultural arenas, a decade whose chief celebrities and cultural heroes were aptly identified by Christopher Booker as "the neophiliacs." More recently, toward the end of the hung-over and gloomy 1970s, George Steiner also indicated the irresponsible capitulation of modern culture generally and of modern philosophy specifically to narrow careerism and its concomitant logic-chopping and esoteric specializations. The results of "desperately fashion-able" academic obsessions such as semiotics are "often of spurious profundity and the jargon that surrounds them is corrupting," Steiner has argued. "What is clear is this: Philosophy, in the current Anglo-American vein, has largely relinquished—indeed it has scorned—those central areas of metaphysics, of ethics, of esthetics, of political thought, that constituted the mainstream and splendor of the philosophic tradition. . . . What it offers instead is dry tack for teeth on edge."[36]

There are of course exceptions to this willfully precipitated decline of wisdom and its accompanying and corroding jargon and cant, and Lewis provides one of them. The Franco-American cultural historian Jacques Barzun, himself a force for clarity and an enemy of cant in the life of his adopted country, where heedless specialist obscurantism has been particularly virulent, wrote in review of a posthumous volume of Lewis's essays on theology and ethics,

> One need not be a believer in Lewis's church to profit from his candor and powers of reasoning on common predic-aments. One of his most telling pieces is on National Repen-tance. Apply its teaching to any of the fashionable emotions and see how many survive. Then, the mind cleared of easy sophistication, start afresh to find out what you think with your whole being about the subjects he proposes to unclut-tered mother wit.[37]

The defense and application of "uncluttered mother wit" is characteristic of almost all of Lewis's writing, but most especially in *The Abolition of Man*, a philosophical essay for the layman which is a kind of survival kit against the furious foolishness and rampant obscurantism brought about by scientistic ideology.

V

The Abolition of Man

The relativization of the absolute leads to the absolutization of the relative.

Sergei Levitzky

Everything except the *sic volo, sic jubeo* has been explained away . . . When all that says 'it is good' has been debunked, what says 'I want' remains. It cannot be exploded or 'seen through' because it never had any pretensions.

C. S. Lewis,
The Abolition of Man

M ODERN scientistic doctrine holds all fact to be objective and all value to be subjective. To call it a "doctrine" is to draw attention to the fact that its characteristic assumption that only factual statements have validity is itself nonfactual, speculative, and dogmatic; it is, in fact, a diabolically ironic article of faith. The worship of scientism is nothing new; it has been with us popularly for well over a century now. In the skewed genius of Nietzsche it received its most tragic and dramatic formulations as it impinged on the realm of ethics, for he argued for nothing less than complete rejection of the *philosophia perennis*. In the absence of that great philosophical tradition with its insistence upon an objective, rationally explicable Good, there is left a great gaping hole in any imaginable intellectual conception of the world, and objective grounds for any and all conduct, including intellectual endeavor, are completely ruled out.

If God is dead, then everything is permitted—Nietzsche and Dostoevsky both drew the same conclusion. Put less dramatically and theologically, it might be said that if there is no Good, then everything is permitted, because there are no grounds for calling anything—an act, a statement, an event, a choice—either good or bad, right or wrong, valuable or worthless, important or unimportant. The end of the moral interpretation of reality, Nietzsche would have it, issues in the aesthetic interpretation of reality: life is an aesthetic phenomenon "beyond good and evil."[1] The only "goods" are subjective and hedonistic—the maximization of pleasure and the minimization of pain by individuals for themselves.

Nietzsche's histrionic evangel was rendered in a more technical and systematic form in the 1930s by the Oxford philosopher A. J. Ayer, and it was to Ayer's version of the creed that Lewis pointedly responded in a number of essays and books,

74

but especially in the novel *That Hideous Strength* and in *The Abolition of Man*. Lewis wrote in the spirit of Whitehead's and Huxley's attacks on scientific materialism but with a somber sense appropriate to the events of the 1930s and 1940s, a sense of history working out an awful judgment on men for contradictory ideas fecklessly flaunted and propagated. Albert Camus attributed the rise of Nazism to a widespread sense of the metaphysical absurdity of life, the view of a bohemian minority in the nineteenth century that gained a far greater currency in the early decades of the twentieth century. The same is true of course of Communism, which is, in addition, as the historian Willi Schlamm put it, "the culminating hubris of Promethean man who reaches out for the world and means to remake creation. It is scientism gone political."[2]

Ideas have consequences, Lewis insisted, and although the problems of modernity have been immensely complicated by technical innovations such as the advent of automation, the factory system, and the massive increases in the speed of communication and transportation, the root of the problem remains philosophical. Lewis's point in *The Abolition of Man* is not simply that the consequences of scientific materialism are bad, but that it is internally inconsistent and false. In his criticism of this heresy he claims no originality beyond that which can be said to derive from remaining faithful to the best that has been thought and said, especially in the tradition of Western philosophy and ethics.

A chief feature of the *philosophia perennis* is the belief that rationality is at least potentially capable of giving us a true picture of reality if we commit ourselves to consistency and non-contradiction. With the Platonists and Scholastics, Ancient and Modern, Lewis argues that reason coherently used leads inevitably to an apprehension of a Good which is no mere human projection but which human rationality is uniquely fitted to infer, recognize, and choose to live by. This Good is the basis not only of morality but of validity itself. Against naturalism and the desperate and contradictory subjectivism and relativism that are its fruits, it is Lewis's obstinate common-sense contention that validity and morality are real and that every rational person acts as if they are, to the extent that he is sane and honorable. Although validity and morality are not identical, they are logically interdependent, and neither can consistently exist without the other; as Anthony Quinton has succinctly said, "valid statements are after all statements that we ought to accept."

Without a doctrine of objective validity, only individual desire remains as a standard to determine action. In the hands of an empowered elite, the capacity to reorder society with the techniques of a vastly powerful and unchecked science is virtually limitless and, of course, open to monstrous misuses—although, Lewis reiterates, the valuation of monstrousness would be irrelevant within an ethical framework based solely on the dictates of personal desire:

> what never claimed objectivity cannot be destroyed by subjectivism. The impulse to scratch when I itch or to pull to pieces when I am inquisitive is immune from the solvent which is fatal to my justice, or honour, or care for posterity. When all that says 'it is good' has been debunked, what says 'I want' remains. It cannot be exploded or 'seen through' because it never had any pretensions. The Conditioners [those scientists—e.g., Behaviorists—bent on redesigning the human species], therefore, must come to be motivated simply by their own pleasure. I am not speaking here of the corrupting influence of power nor expressing the fear that under it our Conditioners will degenerate. The very words *corrupt* and *degenerate* imply a doctrine of value and are therefore meaningless in this context.[3]

The triumph of personal desire over objective validity as a standard of behavior creates what is tantamount to a moral vacuum into which will rush disordered passions bloated in their abnormal freedom from any constraint. Lewis analyzed egotism in terms of three such passions: the *libido sentiendi*, the *libido sciendi*, and the *libido dominandi*—the lust for sensation, the destructive lust for knowledge, and the will to power, respectively. The egotistical will to power and pleasure jumps to supercede any less explicit idea or sentiment of obligation to some criterion or good outside the self. It assumes, as had the Marquis de Sade, that the pleasure of the self is the highest good and that life is an aesthetic phenomenon; under such conditions, society becomes a jungle in which eventually a contest for the survival of the fittest can only favor the Nazi, the Hell's Angel, the robber baron, and the gangster. In the profound formulation of the emigré Russian philosopher Levitzky, "the relativization of the Absolute"—the destruction of the concept of the objective Good—leads necessarily to "the absolutization of the relative"—the belief that the self and its pleasures constitute the only good, the center of the universe of meaning.

But Lewis's complaint extends beyond the monstrous practical implications of this line of reasoning to its essential inconsistency and self-contradiction. It suggests that no validity can be claimed for anything—yet the scientistic modern naturalist inevitably attributes validity overtly or covertly to some beliefs, statements, acts, and procedures. Morality and validity cannot be derived from scientific analysis and empirical knowledge, but systematic and coherent scientific analysis is equally impossible without implicit or explicit reference to a rational doctrine of metaphysical validity. Of course, nature is not the culprit but the victim of this dogmatic reductionism: "I take it," Lewis writes, "that when we understand a thing analytically and then dominate and use it for our own convenience we reduce it to the level of 'Nature' in the sense that we suspend our judgments of value about it, ignore its final cause (if any), and treat it in terms of quantity." This reduction of everything to the spatial and temporal, to the quantitative, is a chosen intellectual attitude, the *libido dominandi*: "We reduce things to mere Nature *in order that* we may 'conquer' them," and the "price of conquest is to treat a thing as mere Nature." The logical outcome of this tendency is to "take the final step of reducing our own species to the level of mere Nature."[4]

The intellectual reduction of man to a natural mechanism has of course been realized by the likes of B. F. Skinner, the behaviorists, and all the other eager conditioners. They have demonstrated tragically that it is, indeed, as Lewis says, "in Man's power to treat himself as a mere 'natural object' and his own judgements of value as raw material for scientific manipulation to alter at will."[5] But as Chesterton has pointed out, this violates common sense and the soul of rationality. We assume in ourselves—experimenters assume in themselves—rational attributes, free will, rational consistency, openness to evidence, desire for truth, and in short, all those nonquantifiable qualities that we rigorously exclude from the human objects of our inspection. The end determines the means, and the means, the end: the desire for power, the *libido dominandi, scientia potestas est*, exceeds the love of truth, and scientism attacks the canons and doctrines of philosophical truth from which it derives whatever rational consistency and validity it maintains.

The traditional values and ideas that Lewis defends against this imperial scientism are disinterestedness, truth, and the distinction between ends and means. "In coming to understand anything," he writes in *An Experiment in Criticism*, "we are rejecting the facts as they are *for us* in favour of the facts as they are."[6] This is the age-old intellectual and philosophical task of trying to see things in themselves as they really are, to know the truth first, not merely to know "things" as they provide useful means and instruments to the satisfaction of our personal or collective desires and the maximization of our pleasures. Lewis argues that we are obligated to transcend self-interest in order to know the truth and that the pursuit of our own pleasure ought to be restricted by and subordinated to the truth; he would say, with Lara in *Dr. Zhivago*, that "God put me on earth to call things by their right names." This is even more true of the metaphysical realm than of the physical. "For the wise men of old," Lewis wrote "the cardinal problem had been how to conform the soul to reality, and the solution had been knowledge, self-discipline, and virtue." But for the practitioners of applied science—for those who see through scientistic spectacles—"the problem is how to subdue reality to the wishes of men: the solution is a technique."[7]

With the growth of scientism has come a massive increase in the powers of technology and applied science to change and manipulate not only the physical landscape but the mental and human landscape too. As the means and instruments proliferate, the distinction between ends and means seems to grow more obscure in modern culture, so much so that finally man himself can be seen as a means to undetermined ends; he is deluded by what William Barrett calls "the illusion of technique." There is no longer any question of "conforming the soul to reality"; there is only the question of increasing our power over and pleasure in a world of objects.

Lewis soldiered on with the *philosophia perennis* in days and years when it was unfashionable in the extreme to do so. He luminously exemplified the long tradition of *sapientia* and the critique of scientism that we have sketched. Bertrand Russell (intermittently), Martin Heidegger, and Jacques Ellul have also criticized the lust for "power-knowledge," which Nietzsche had called the "will to power" and the scholastics had called the *libido dominandi*, as the central feature of modern life. This insatiable lust grows out of the anxiety of the isolated self in an absurd world that is the consensus of reality bequeathed to modern man by scientific materialism.

Lewis attacked this consensus of absurdity on the grounds that it omits the "one thing needful," the intelligible Good which alone makes purposeful sense of intellect and morality. He saw that one of the essential forms of idolatry was the quest for power without goodness—the modern totem of technical power which he called "that hideous strength." It is the idol created by the will to power "liberated" from that intelligible Good which is its only true curb, guide, and governor. He argued that the refusal to know this Good is caused by cankered wills rather than by weak minds and that it is the duty of the wise man to work to restore its lustre. He held self-knowledge and *sapientia* primary, and the knowledge of objects secondary, and warned that if the latter took precedence over the former, error and disaster would not lag far behind.

A traditional rational Theist, Lewis was no Manichean or radical transcendentalist: he did not doubt either the reality or the goodness of the natural world; he merely insisted that "second things suffer by being put first"—that nature was both real and good, but neither the sole nor the superior reality. "The attempt," he wrote in *Letters to Malcolm*, "is not to escape from space and time and from my creaturely situation as a subject facing objects. It is more modest: to reawaken the awareness of that situation."[8] An awareness of others as the subjects as well as the objects of our consciousness and a recognition of the distinction between man and thing wherein man is essentially *no thing* at all, but a value, *res sacra homo*, an ultimate end, like the God in whose image he is made—this is traditional metaphysical *sapientia*, this is wisdom.

The danger of mechanized civilization, of an industrial and commercial society, is that it tends to degrade the distinctive nature of man by emphasizing speed, size, quantity, and the maximization of physical pleasure by applied technical power at the expense of more crucial human values. The catastrophically dehumanizing means by which the Industrial Revolution came into being was as apparent to Lewis as it had been to the great Victorian social critics before him. As the Hammonds put it, "in adapting this new [industrial] power to the satisfaction of its wants England could not escape from the moral atmosphere of the slave trade: the atmosphere in which it was the fashion to think of men as things."[9] Lewis made a similar point in his inaugural address at Cambridge, *"De Descriptione Temporum"*: "The birth of the machines," he said, "really is the greatest change in the history of Western man."[10]

The danger that Lewis detected is not only, as Chesterton put it, that many of the labor-saving devices "save labor a good deal more than they ought," by destroying the diversity of crafts and traditions, but that they discourage the expression of specifically human skills and qualities; they breed passivity in response to experience: television discourages reading and thus literacy, the telephone discourages letter-writing, the automobile discourages walking, the contraceptive discourages chastity, planned obsolescence discourages thrift and permanence. The underlying threat of commercial civilization lies in its emphasis on the sensate and the ephemeral over the transcendental and the lasting. It encourages an obsession with objects and quantities instead of subjects and qualities, as witness the pejorative connotations that the popular media have helped to attach to such words as *discrimination.*

Lewis insisted that man was not a thing, but an essence, a soul, and that it ill profits a man to gain the whole material world at the expense of the elementary self-knowledge that tells him that he is a soul qualitatively distinct from and superior to those things. Persons are ultimate ends and ought never to be treated only as means; they always have the character of "thou" and ought never to be treated merely as "it." This is the root of common sense dualism which Lewis, like Chesterton, considered the birthright of *homo sapiens*, without which men would inevitably turn first into *homo sciens* and then into mere nature. He placed these truths at the core of civil humanity, the *res publica* on which law, manners, and civilization itself are built; they are the roots from which stem human virtue, decency, accomplishment, and sanity itself. He did not argue that they constituted a panacea: they are merely the truth, the only path an honest person ought to tread.

That Lewis understood well his own position as a moralist is apparent from what he had to say about others so called. In *The Screwtape Letters*, Screwtape writes to his young protegé that "all the great moralists are sent by the Enemy [viz., God] not to inform men, but to remind them, to restate the primeval moral platitudes against our continual concealment of them."[11] In his great scholarly book on medieval English literature, *The Allegory of Love*, Lewis writes of Langland that as a moralist he has nothing original or unique about him; in fact, he says, "it is doubtful whether any moralist of unquestioned greatness has ever attempted more (or less) than the defence of the universally acknowledged; for 'men more frequently required to be reminded than informed'."[12] He sought only to add his voice to that of Johnson and Hooker and Aquinas and Aristotle and Plato, in defense of the ancient orthodox tradition of ethics, the great central metaphysical tradition, the *philosophia perennis*. In distinguishing what Pope called "solid worth" from "empty show," our century has no more reliable or entertaining guide than C. S. Lewis. He bowed to the constraint of truth and thus climbed its ladder.

Notes

Chapter One

1. Chad Walsh, "Impact on America," in *Light on C. S. Lewis*, ed. Jocelyn Gibb (London: Geoffrey Bles, 1965), p. 106.

2. Richard Ellman, *James Joyce* (1959; New York: Oxford Univ. Press, 1965), pp. 64-65.

3. Willey, *Christianity Past and Present* (Cambridge: Cambridge Univ. Press, 1952), p. 55.

4. Paul Roche, quoted by Russell Kirk, *Confessions of a Bohemian Tory* (New York: Fleet, 1963), p. 233.

5. Al-Ghazali, quoted by P. L. Berger, *The Heretical Imperative* (Garden City, New York: Anchor-Doubleday, 1979), pp. 90-91.

6. Hooker, *The Works of Richard Hooker* (London: Oxford Univ. Press, 1850), I, 172.

7. Pope's comment appears in the Preface to the 1717 edition of his *Complete Works*, rpt. in the Twickenham Edition of the *Poems of Alexander Pope*, ed. E. Audra and Aubrey Williams (New Haven: Yale Univ. Press, 1961), I, 7.

8. Lewis, *Studies in Words* (Cambridge: Cambridge Univ. Press, 1967), p. 150. "Nothing," Lionel Trilling has said, "is more characteristic of modern literature than its . . . canonization of the primal, non-ethical energies." *Beyond Culture* (New York: Viking Press, 1968), p. 19.

9. Lewis, "The Poison of Subjectivism," in *Christian Reflections*, ed. Walter Hooper (Grand Rapids, Michigan: Eerdmans, 1967), p. 78.

10. William Barrett, *The Illusion of Technique: A Search for Meaning in a Technological Civilization* (Garden City, New York: Anchor-Doubleday, 1978), pp. 232, 233.

11. Lewis, *The Abolition of Man* (1943; New York: Macmillan, 1960), p. 47.

12. Sylvestris, quoted in *The Metalogicon of John of Salisbury: A Twelfth-Century Defense of the Verbal and Logical Arts of the Trivium*, trans. Daniel D. McGarry (Berkeley: Univ. of California Press, 1955), p. 167.

13. Lewis, in a letter to *The Christian Century,* 31 Dec. 1958; rpt. in *God in the Dock: Essays in Theology and Ethics*, ed. Walter Hooper (Grand Rapids, Michigan: Eerdmans, 1970), p. 338.

14. Swift, *Satires and Personal Writings*, ed. W. A. Eddy (London: Oxford Univ. Press, 1932), pp. 274-75.

15. Lewis, quoted by Christopher Derrick, "C. S. Lewis: Against the Cult of Culture," *The Times* (London), 28 Apr. 1973.

16. G. M. Young, *Victorian England: Portrait of an Age* (London: Oxford Univ. Press, 1953), p. 160.

17. Huxley, *Ends and Means* (London: Chatto & Windus, 1937), p. 276.

18. The quotation serves as epigraph to Book Three.

19. Lewis, *A Preface to Paradise Lost* (London: Oxford Univ. Press, 1942), p. 55.

20. Dostoevsky, quoted by V. Borisov in *From Under the Rubble* (Boston: Little, Brown, 1975), p. 202.

21. Lewis, *A Preface to Paradise Lost*, p. 63.

22. Miller, in *Diacritics*, 2 (Winter 1972), 6.

23. Lewis, "Lilies That Fester," in *The World's Last Night and Other Essays* (New York: Harcourt, Brace, 1960), p. 39.

24. Chesterton, "The New Groove," in *The Common Man* (London: Sheed and Ward, 1950), p. 112.

25. Lewis, "Religion: Reality or Substitute?", in *Christian Reflections*, p. 41.

26. Lewis, *The Screwtape Letters* (1941; New York: Macmillan, 1974), pp. 107-08.

27. A modern Chinese translation of St. John's Gospel by Dr. John Wu uses the word *Tao* to capture the sense of the universality of Rational Good implicit in the Greek *logos*. For Lewis's discussion of his use of the word, see *The Abolition of Man*, especially pp. 17 ff., and also *A Mind Awake: An Anthology of C. S. Lewis*, ed. Clyde S. Kilby (London: Geoffrey Bles, 1968), p. 36.

28. Lewis, "Edmund Spenser," in *Major British Writers*, ed. G. B. Harrison (New York: Harcourt, Brace, 1954), I, 99-100.

29. Griffiths, "The Adventure of Faith," in *C. S. Lewis at the Breakfast Table and Other Reminiscences*, ed. James T. Como (New York: Macmillan, 1979), p. 24.

30. Lewis, quoted by Leo Baker, "Nearing the Beginning," in *C. S. Lewis at the Breakfast Table*, p. 10.

31. Lewis, quoted by Paul L. Holmer, *C. S. Lewis: The Shape of His Faith and Thought* (1976; London: Sheldon Press, 1977), p. 47.

32. Fish, *Surprised by Sin: The Reader in Paradise Lost* (London: Macmillan, 1967), p. 90, n. 1.

33. Bronowski, *The Identity of Man* (Garden City: Natural History Press, 1965), p. 2.

34. Chesterton, *The Everlasting Man* (New York: Dodd, Mead, 1925), p. 20.

35. Lewis, "Religion Without Dogma?", in *God in the Dock*, p. 138.

36. Lewis, "'Bulverism'; or, The Foundation of Twentieth-Century Thought," in *God in the Dock*, p. 276.

37. Blake, "Jerusalem," ch. 2, pl. 42, ll. 25-26, in *The Complete Writings of William Blake*, ed. Geoffrey Keynes (New York: Random House, 1957), p. 670.

38. Stanley L. Jaki, *The Origin of Science and the Science of Its Origin* (Edinburgh: Scottish Academic Press, 1978), p. 21.

39. Whitehead, quoted by William Barrett, *The Illusion of Technique*, p. 335.

40. Lewis, *"De Futilitate,"* in *Christian Reflections*, p. 65.

41. Richard Poirier, quoted by M. J. Sobran, "Lay-deez and Gentlemeen: Pree-zen-ting . . . the Orgulous . . . Supersubtle . . . Garry . . . WILLS!!!", *National Review*, 22 June 1973, p. 681.

42. Sobran, *National Review*, 22 June 1973, p. 681. "To Johnson the great truths are above all *public,* available equally to all reflective men. . . . When he explains something, however magisterially, it is like the sharing of a common possession, a communion—exactly comparable to the breaking of bread."

43. Speaight, quoted by R. L. Green and Walter Hooper, *C. S. Lewis: A Biography* (1974; London: Collins Fount, 1979), p. 210.

44. Lewis, quoted in *C. S. Lewis: A Biography*, p. 54.

45. Lewis, quoted in *C. S. Lewis: A Biography*, p. 105.

46. Griffiths, "The Adventure of Faith," in *C. S. Lewis at the Breakfast Table*, p. 12.

47. Ibid, p. 19.

48. Ibid., p. 17.

49. Lewis, "The Weight of Glory," in *The Weight of Glory and Other Addresses* (Grand Rapids, Michigan: Eerdmans, 1965), p. 38.

50. Lewis, "Equality," *The Spectator*, 27 Aug. 1943, p. 92.

51. Gross, *The Rise and Fall of the Man of Letters* (London: Weidenfeld and Nicolson, 1969), p. 220.

52. Lewis, quoted in *C. S. Lewis: A Biography*, p. 204.

Chapter Two

1. Rabelais, quoted by Harry Levin in *The Overreacher: A Study of Christopher Marlowe*; rpt. in *Marlowe's Dr. Faustus: A Selection of Critical Essays*, ed. J. D. Jump (London: Macmillan, 1969), p. 134.

2. *Tuttolibri*, Weekly Book Review of *La Stampa* (Turin, Italy), 8 Apr. 1978, p. 16.

3. Hailsham, *The Door Wherein I Went* (1975; London: Collins Fount, 1978), p. 4.

4. Lewis, *Perelandra* (New York: Macmillan, 1965), p. 81.

5. Penfield, *The Mystery of the Mind* (Princeton: Princeton Univ. Press, 1975), p. 80.

6. Chadwick, *The Secularization of the European Mind in the Nineteenth Century* (Cambridge: Cambridge Univ. Press, 1975), p. 164.

7. Whitehead, quoted by Joseph A. Mazzeo, *The Theory and Practice of Interpretation: A Commonplace Book* (New York: privately printed, 1975), p. 65.

8. Jaki, *The Origin of Science and the Science of Its Origin* (Edinburgh: Scottish Academic Press, 1978), p. 103.

9. See *Leibniz Selections*, ed. P. P. Wiener (New York: Scribner's, 1951), p. 70.

10. Lewis, *The Abolition of Man* (1943; New York: Macmillan, 1960), p. 48.

11. James Franck, quoted by Jaki, *The Origin of Science*, p. 153, n. 54.

12. Donne, "Of the Progress of the Soul: The Second Anniversary,"

ll. 291-95, in *Donne: The Complete Poetry and Selected Prose*, ed. C. M. Coffin (New York: Modern Library, 1952), p. 206. (Slightly modernized.)

13. Bacon, quoted by Simone Weil, *Lectures in Philosophy*, trans. Hugh Price (Cambridge: Cambridge Univ. Press, 1978), p. 214.

14. Bacon, quoted by William Barrett, "Heidegger and Modern Existentialism: Dialogue with William Barrett," in *Men of Ideas*, ed. Bryan Magee (New York: Viking Press, 1978), p. 87.

15. Swift, *Gulliver's Travels* (1726; London: J. M. Dent, 1940), p. 198.

16. Jaki, *The Origin of Science*, pp. 25-26.

17. Johnson, "Milton," in *Lives of the English Poets* (1779; London: Oxford Univ. Press, 1906), I, 73.

18. Cicero, quoted by Walter Jackson Bate, *From Classic to Romantic: Premises of Taste in Eighteenth-Century England* (Cambridge: Harvard Univ. Press, 1946), p. 20.

19. Johnson, "On the Isle of Skye," in *Johnson: Complete English Poems*, ed. J. D. Fleeman (Baltimore: Penguin, 1971).

20. Chateaubriand, *An Historical, Political, and Moral Essay on Revolutions, Ancient and Modern* (London: Henry Colburn, 1815), p. 357.

21. Toynbee, Introd., *The Gods of Revolution*, by Christopher Dawson (New York: New York Univ. Press, 1972), p. x.

22. Trilling, "Mind in the Modern World," in *The Last Decade: Essays and Reviews, 1965-1975* (New York: Harcourt, Brace, 1978), p. 107.

23. Heller, quoted by Daniel Bell, *The Cultural Contradictions of Capitalism* (1976; New York: Basic Books, 1978), pp. 160-61.

24. Goethe, quoted by Erich Heller, "Goethe," in *Atlantic Brief Lives: A Biographical Companion to the Arts*, ed. L. Kronenberger (Boston: Little, Brown, 1965), pp. 324-25.

25. Coleridge, quoted by Basil Willey, *Samuel Taylor Coleridge* (New York: W. W. Norton, 1972), p. 13.

26. Coleridge, quoted in *Coleridge*, p. 100.

27. Coleridge, quoted in *Coleridge*, p. 177.

28. Willey, *Coleridge*, p. 175.

29. Kierkegaard, "The Present Age," in *The Living Thoughts of Kierkegaard*, ed. W. H. Auden (New York: David McKay, 1952), p. 37.

30. Ibid., p. 39.

31. Kierkegaard, quoted by Philip Rieff, *Fellow Teachers* (1972; New York: Dell-Delta, 1975), pp. 158-59.

32. Kierkegaard, *The Journals*, trans. Alexander Dru (London: Oxford Univ. Press, 1938), p. 181.

33. Kierkegaard, quoted by John Passmore, *Science and Its Critics* (London: Duckworth, 1978), p. 2.

34. Nietzsche, *The Gay Science*, in *A Nietzsche Reader*, trans. R. J. Hollingdale (Harmondsworth, England: Penguin, 1977), p. 158.

35. Melville, from *Clarel*, "Ungar and Rolfe," in *The Oxford Book of American Verse*, ed. F. O. Matthiessen (New York: Oxford Univ. Press, 1950), p. 404.

36. Emerson, from "Ode: Inscribed to W. H. Channing," in *Selections from Ralph Waldo Emerson*, ed. Stephen E. Whicher (Boston: Houghton Mifflin, 1957), p. 440.

37. Emerson, quoted by Trilling, *The Last Decade*, p. 121.

38. Vogt and Feuerbach are quoted by Owen Chadwick, *The Secularization of the European Mind*, pp. 166, 169.

39. Disraeli, quoted by John Passmore, *A Hundred Years of Philosophy* (New York: Basic Books, 1966), p. 36.

40. Arnold, quoted by Andrew Boyle, *Climate of Treason* (London: Coronet, 1980), p. 21.

41. Newman, quoted by M. J. Svaglic, Introd., *The Idea of a University*, by John Henry Newman (New York: Holt, Rinehart, 1960), p. xxvi.

42. Newman, *Idea of a University*, p. 55.

43. Ibid., pp. 52-53.

44. Newman, quoted by Svaglic, Introd., *Idea of a University*, p. xxi.

45. Newman, *Idea of a University*, pp. 72-73.

46. Chadwick, "In Pursuit of Excellence," rev. of *The Idea of a University*, by John Henry Newman, *Times Literary Supplement*, 13 Aug. 1976, p. 1002.

47. Martin, "Order and Rule: A Critique of Spontaneity," in *Tracts Against the Times* (London: Lutterworth, 1973), p. 147.

48. Ruskin, *The Genius of John Ruskin: Selections from His Writings*, ed. John D. Rosenberg (New York: Braziller, 1963), pp. 437, 444, 230, 231.

49. John Passmore, *The Perfectibility of Man* (New York: Scribner's, 1970), p. 250.

50. Said, *Orientalism* (New York: Pantheon, 1978), p. 140.

51. Dostoevsky, *The Possessed*, trans. A. R. MacAndrew (New York: New American Library, 1962), p. 237.

52. Orwell, "The Prevention of Literature," in *The Collected Essays of George Orwell*, ed. S. Orwell and I. Angus (New York: Harcourt, Brace, 1968), IV, 70.

53. Passmore, *Science and Its Critics*, p. 89.

54. In the *Chesterton Review*, 3 (1977), 314.

55. Chesterton, *Eugenics and Others Evils* (1922; New York: Dodd, Mead, 1927), p. 98.

56. Sinsheimer, quoted in *The People Shapers* (London: Futura Publications, 1978), p. 230.

57. Chesterton, *The Spice of Life* (Beaconsfield, England: D. Finlayson, 1964), p. 43.

58. Chesterton, *Generally Speaking* (London: Methuen, 1937), p. 239.

59. Ibid., p. 123.

60. Chesterton, *The Common Man* (London: Sheed and Ward, 1950), p. 116.

61. Ibid., pp. 116-17.

62. Chesterton, "Shaw the Dramatist," in *G. K. Chesterton: A Selection from His Non-Fictional Prose*, ed. W. H. Auden (London: Faber and Faber, 1970), p. 111.

63. Chesterton, in a letter to the *Daily News* dated 3 Feb. 1906; rpt. in *The Man Who Was Orthodox*, ed. A. L. Maycock (London: Dobson, 1963), pp. 155-56.

64. Chesterton, *Generally Speaking*, p. 12.

65. Chesterton, *All Things Considered* (London: Methuen, 1908), p. 187.

66. Chesterton, *Generally Speaking*, p. 90.

67. Ibid., pp. 88-89.

68. Acton, quoted by Herbert Butterfield, *The Whig Interpretation of History* (1931; London: G. Bell, 1951), p. 112.

69. See William Barrett, *The Illusion of Technique: A Search for Meaning in a Technological Civilization* (Garden City, New York: Anchor-Doubleday, 1978).

70. Jacques Ellul, quoted in *Worldview*, Dec. 1980, p. 31.

71. Chesterton, *The Thing* (London: Sheed and Ward, 1929), p. 13.

72. Chesterton, *Heretics* (New York: John Lane, 1905), p. 53.

73. Williams, quoted by Christopher Derrick, *The Delicate Creation: Towards a Theology of the Environment* (London: Tom Stacey, 1972), p. 18.

74. Huxley, *Ends and Means* (London: Chatto and Windus, 1969), p. 252.

75. Chesterton, *The Common Man*, p. 173.

76. Marcel, *The Decline of Wisdom* (London: Harvill, 1954), p. 49.

77. Chesterton, *St. Thomas Aquinas* (London: Sheed and Ward, 1933), p. 71.

Chapter Three

1. Chesterton, *The Common Man* (London: Sheed and Ward, 1950), p. 9.

2. Gingerich, *Christian Herald*, Dec. 1978, p. 34.

3. Jastrow, *New York Times Magazine*, 25 June 1978, p. 26.

4. Kroner, *Culture and Faith* (Chicago: Univ. of Chicago Press, 1951), p. 11. Kroner also makes the point that "[p]eople begin to adore science after science has deprived them of their proper object of adoration, because they need such an object and are fond of adoration" (p. 112).

5. Skolimowski, quoted by Colin Pritchard, "Science, Faith and the Vision of a New Society," *Theology*, 80 (1977), 333.

6. Phenix, *Teachers College Record*, 57 (Oct. 1955), 30.

7. Schumacher, *A Guide for the Perplexed* (New York: Harper & Row, 1977), pp. 4-5.

8. Roszak, "The Monster and the Titan: Science, Knowledge, and

Gnosis," *Daedalus*, 103 (Summer 1974), 26.

9. As, for instance, in *From Under the Rubble* (1974; Boston: Little, Brown, 1975).

10. See Milovan Djilas, *The New Class: An Analysis of the Communist System* (1954; New York: Praeger, 1957).

11. Martin, *The Religious and the Secular: Studies in Secularization* (New York: Schocken, 1969), p. 37.

12. Whitehead, *Adventures of Ideas* (1933; Cambridge: Cambridge Univ. Press, 1961), p. 37.

13. Chesterton, *Orthodoxy* (London: Bodley Head, 1908), pp. 211-12.

14. Whitehead, *Adventures of Ideas*, pp. 228-29.

15. Bridgman, "The Way Things Are," in *The Limits of Language*, ed. Walker Gibson (New York: Hill and Wang, 1962), pp. 42, 44.

16. Born, quoted in *The Reader's Adviser*, ed. Winifred F. Courtney, 11th ed. (New York: Bowker, 1969), II, 434.

17. Jonas, *The Phenomenon of Man: Toward a Philosophical Biology* (New York: Harper and Row, 1966), p. 196.

Chapter Four

1. Collingwood, *Speculum Mentis; or, The Map of Knowledge* (Oxford: Clarendon, 1924), p. 281.

2. Ayer, in a dialogue with Bryan Magee, *Men of Ideas* (New York: Viking Press, 1978), pp. 130, 131.

3. Barrett, *The Illusion of Technique: A Search for Meaning in a Technological Civilization* (Garden City, New York: Anchor-Doubleday, 1978), pp. 211-12.

4. Ellul, quoted by Robert Angus Buchanan, "Technology, Conceptions of," *New Encyclopaedia Britannica: Macropaedia*, 1974 ed.

5. Lewis, quoted by R. L. Green and Walter Hooper, *C. S. Lewis: A Biography* (1974; London: Collins Fount, 1979), p. 173.

6. Arnold, *Culture and Anarchy* (London: Smith, Elder, 1869), pp. 15, x.

7. Ibid., p. 15.

8. Leavis, Afterword, *The Pilgrim's Progress*, by John Bunyan (New York: New American Library, 1964), p. 292.

9. Leavis, *The Common Pursuit* (London: Chatto and Windus, 1952), p. 87.

10. Leavis, quoted in *Johnsonian Studies*, ed. M. Wahba (Cairo: n.p., 1962), p. 29.

11. Leavis, quoted in Reuters Obituary, Rome *Daily American*, 2 May 1978, p. 6.

12. Leavis, *Lectures in America* (London: Chatto and Windus, 1969), p. 51.

13. Leavis, quoted by J. Harvey, "F. R. Leavis: An Appreciation,"

Encounter, 52 (May 1979), 65.

14. Leavis, quoted in the *New York Review of Books*, 16 Aug. 1979, p. 31.

15. Roger Poole, *London Review of Books*, 20 Dec. 1979, p. 8.

16. Lewis, quoted by Stanley E. Fish, *Self-Consuming Artifacts: The Experience of Seventeeth-Century Literature* (Los Angeles: Univ. of California Press, 1972), p. 410.

17. Newman [Catholicus], *The Tamworth Reading Room* (London: John Mortimer, 1841), p. 19.

18. Sutherland, *The English Critic* (London: H. K. Lewis, 1952), pp. 14-15.

19. Lewis, "Christianity and Culture," in *Christian Reflections*, ed. Walter Hooper (Grand Rapids, Michigan: Eerdmans, 1967), p. 12.

20. Amis, "Beer and Beowulf," rev. of *The Inklings*, by Humphrey Carpenter, *New Statesman*, 20 Oct. 1978, p. 510.

21. Lewis, quoted by John Wain, "A Great Clerke," in *C. S. Lewis at the Breakfast Table and Other Reminiscences*, ed. James T. Como (New York: Macmillan, 1979), p. 73.

22. Eliot, *Poetry and Drama* (London: Faber and Faber, 1951), p. 35.

23. Leavis, quoted by Lionel Trilling, "Dr. Leavis and the Moral Tradition," in *A Gathering of Fugitives* (Boston: Beacon, 1956), p. 102.

24. Lewis, *An Experiment in Criticism* (Cambridge: Cambridge Univ. Press, 1961), pp. 136-37, 124-25.

25. Ibid., p. 119.

26. Lewis, *Studies in Words* (Cambridge: Cambridge Univ. Press, 1967), p. 7.

27. In "The Tutor: A Portrait," in *C. S. Lewis at the Breakfast Table*, pp. 63-64, Derek Brewer reports that he once asked Lewis how he got along with Leavis at Cambridge: "He replied that they were always on opposing sides of the question in any committee on which they met, but that Leavis 'was all right,' he was 'saved.' He did not mean religiously, but that ultimately his values were the right ones."

28. Arnold, quoted by Basil Willey, *Nineteenth Century Studies* (New York: Columbia Univ. Press, 1945), pp. 270, 269.

29. Arnold, *Culture and Anarchy*, pp. xxxii, lvi.

30. Lewis, *Experiment in Criticism*, p. 138.

31. Babbitt, *Literature and the American College: Essays in Defense of the Humanities* (1908; Chicago: Regnery Gateway, 1956), p. 13.

32. Lewis, *Experiment in Criticism*, p. 141.

33. Lewis, "The Poison of Subjectivism," in *Christian Reflections*, p. 78.

34. Joseph A. Mazzeo, "Interpretation, Humanistic Culture, and Cultural Change," *Thought: A Review of Culture and Idea*, 51 (March 1976), 81.

35. Maritain, *The Peasant of the Garonne* (New York: Holt, Rinehart, 1968), p. 14.

36. Steiner, "The Lollipopping of the West," *New York Times*, 9 Dec. 1977, p. A27.

37. Barzun, quoted by Eugene McGovern, "Our Need for Such a Guide," *C. S. Lewis at the Breakfast Table*, p. 130.

Chapter Five

1. Nietzsche, *The Birth of Tragedy and The Genealogy of Morals*, trans. Francis Golffing (Garden City, New York: Doubleday, 1956), pp. 42, 10.

2. Schlamm, quoted by G. H. Nash, *The Conservative Intellectual Movement in America Since 1945* (New York: Basic Books, 1976), p. 269.

3. Lewis, *The Abolition of Man* (New York: Macmillan, 1960), p. 41.

4. Ibid., pp. 44-45.

5. Ibid., p. 45.

6. Lewis, *An Experiment in Criticism* (Cambridge: Cambridge Univ. Press, 1961), p. 138. Italics mine.

7. Lewis, *Abolition of Man*, p. 48.

8. Lewis, *Letters to Malcolm* (New York: Harcourt, Brace, 1964), p. 81.

9. John L. and Barbara Hammond, "The Industrial Revolution: The Rulers and the Masses," in *The Industrial Revolution in Britain: Triumph or Disaster?*, ed. Philip A. M. Taylor (Boston: D. C. Heath, 1958), p. 34.

10. Lewis, *"De Descriptione Temporum,"* in *A Mind Awake: An Anthology of C. S. Lewis*, ed. Clyde S. Kilby (London: Geoffrey Bles, 1968), p. 219.

11. Lewis, *The Screwtape Letters* (1943; New York: Macmillan, 1962), p. 107.

12. Lewis, *The Allegory of Love* (1936; London: Oxford Univ. Press, 1958), pp. 158-59.

INDEX OF NAMES

INDEX OF NAMES

R

Rabelais, F   19, 31, 43
Renan, E.   36
Roche, P.   2
Roszak, T.   50
Rousseau   28
Ruskin   28, 36
Russell, B.   28, 79

S

Sade   28, 76
Said, E.   36
St. John   3, 34
Schlamm, W.   75
Schumacher, E. F   50
Shakespeare   30, 39
Sidney   8, 9
Sinsheimer, R. L.   40
Sitwells   64
Skinner, B. F   4, 27, 48, 77
Skolimowski, H.   49-50
Snow, C. P.   18, 57, 62, 64
Sobran, J.   13
Socrates   26, 30
Solzhenitsyn   38, 51
Speaight, R.   14
Spenser   10, 22, 65
Staël, Mme. de   43
Steiner, G.   72
Stoppard, T.   60

Strachey, J.   51
Sutherland, J.   65
Swift   5, 14, 18, 24, 30, 39

T

Tennyson   42
Thucydides   7
Toynbee, A.   27, 37, 44
Trilling, L.   8, 28, 67

V

Vincent of Lerins   3
Virgil   66
Vogt, K.   33

W

Walsh, C.   2
Waugh, E.   60
West, R.   52
Whitehead, A. N.   13, 19, 21, 44, 48, 52, 53, 69, 75
Willey, B.   2, 19, 30, 59
Williams, C.   44
Woolf, V.   64
Wordsworth   28

Y

Young, G. M.   6